FIFTY THINGS THAT MADE THE MODERN ECONOMY

Tim Harford

ABACUS

First published in Great Britain in 2017 by Little, Brown
This paperback edition published in 2018 by Abacus

9 10 8

A CIP catalogue record for this book
is available from the British Library.

Hardback ISBN 978-0-349-14263-0

Typeset in Bembo by M Rules
Printed and bound in Great Britain by
Clays Ltd, Elcograf S.p.A.

Papers used by Abacus are from well-managed forests
and other responsible sources.

Abacus
An imprint of
Little, Brown Book Group
Carmelite House
50 Victoria Embankment
London EC4Y 0DZ

An Hachette UK Company
www.hachette.co.uk

www.littlebrown.co.uk

Tim Harford is a senior columnist at the *Financial Times*, and his writing has appeared everywhere from *Esquire* and *Wired* to the *Washington Post* and the *New York Times*. His previous books include *Adapt*, *The Logic of Life* and the million-selling *The Undercover Economist*. Harford is a visiting fellow at Nuffield College, Oxford, and was the winner of the 2006 Bastiat Prize for economic journalism. On BBC Radio 4, Harford presents *Pop Up Economics* and *More or Less*, which was commended by the Royal Statistical Society for excellence in journalism three years running. He has spoken about his ideas at TED and at the Sydney Opera House.

'Short chapters are a delight in this frenetic age . . . Best of all, the book is constantly surprising. It brims with innovations I didn't know about, as well as ones I thought I knew about but did not'
The Times

'Packed with fascinating detail . . . Harford has an engagingly wry style and his book is a superb introduction to some of the most vital products of human ingenuity' *Sunday Times*

'Harford's richness of detail bespeaks skill both as an economic analyst and as a popular commentator. His sections on barbed wire, passports, the contraceptive pill, infant formula, the bar code and even that IKEA staple, the Billy bookcase, are well researched, racily written and genuinely thought-provoking. His five-page essay on the pill is infinitely subtler (and more feminist) than the usual stuff about empowerment . . . This is an entertaining book that might distract you from your gramophone for more than an evening and will find a secure place beside Harford's other books on your Billy bookcase'
Times Literary Supplement

Contents

To Andrew Wright

1

The Plough

Imagine catastrophe.

The end of civilisation. This complex, intricate modern world of ours is finished. Don't worry about why. Maybe it was swine flu or nuclear war, killer robots or the zombie apocalypse. And now imagine that you – lucky you – are one of the few survivors. You have no phone. Who would you phone anyway? No internet. No electricity. No fuel.

Four decades ago, the science historian James Burke posed that scenario in his TV series *Connections*. And he asked a simple question: surrounded by the wreckage of modernity, without access to the lifeblood of modern technology, where do you start again? What do you need to keep yourself – and the embers of civilisation – alive?

And his answer was a simple yet transformative technology. It's a plough. And that's appropriate, because it was the plough that kickstarted civilisation in the first place. The plough, ultimately, made our modern economy possible. And by doing that, it made modern life possible too, with all its conveniences and frustrations: the satisfaction of good, plentiful food; the ease of a quick web search; the blessing of clean, safe water; the fun of a video game; but also the pollution of air and water; the scheming of fraudsters; and the grind of a tedious job – or no job at all.

Twelve thousand years ago, humans were almost entirely nomadic, hunting and foraging their way into every niche they could find all round the world. But at the time the world was emerging from a cold snap: things were starting to get hotter and dryer. People who had been hunting and foraging in the hills and high plains found that the plants and the animals around them were dying. Animals were migrating to the river valleys in search of water, and people followed. This shift happened in many places and at different times – over eleven thousand years ago in Western Eurasia, nearly ten thousand years ago in India and China, and more than eight thousand years ago in Mesoamerica and the Andes. Eventually it happened almost everywhere.

These fertile but geographically limited river valleys changed the way people got enough to eat: it was less rewarding to roam around foraging for food, but more rewarding to give the local plants some encouragement. That meant breaking up the surface of the soil, which brought nutrients to the surface and let moisture seep deeper, out of sight of the harsh sun. At first they used sharp sticks, held in the hand, but soon they switched to a simple scratching plough, pulled by a pair of cows. It worked remarkably well.

Agriculture began in earnest. It was no longer just a desperate alternative to the dying nomadic lifestyle, but a source of real prosperity. When farming was well established – two thousand years ago in Imperial Rome, nine hundred years ago in Song-dynasty China – these farmers were five or six times more productive than the foragers they had replaced.

Think about that: it becomes possible for a fifth of a society's population to grow enough food to feed everyone. What do the other four-fifths do? Well, they're freed up to special-ise in other things: baking bread, firing bricks, felling trees, building houses, mining ore, smelting metals, constructing roads – in other words, making cities, building civilisation.

But there's a paradox: more abundance can lead to more competition. If ordinary people live at subsistence levels, pow-erful people can't really take much away from them – not if they want to come back and take more the next time there's a harvest. But the more ordinary people are able to produce, the more powerful people can confiscate. Agricultural abundance creates rulers and ruled, masters and servants, and inequality of wealth unheard of in hunter-gatherer societies. It enables the rise of kings and soldiers, bureaucrats and priests – to organ-ise wisely, or live idly off the work of others. Early farming societies could be astonishingly unequal. The Roman Empire, for example, seems to have been close to the biological limits of inequality: if the rich had had any more of the Empire's resources, most people would simply have starved.

But the plough did more than create the underpinning of civilisation – with all its benefits and inequities. Different types of plough led to different types of civilisation.

The first simple scratch ploughs used in the Middle East worked very well for thousands of years – and then spread west to the Mediterranean, where they were ideal tools for cultivating the dry, gravelly soils. But then a very different tool, the mouldboard plough, was developed – first in China more than two thousand years ago, and much later in Europe. The mouldboard plough cuts a long, thick ribbon of soil and turns it upside down. In dry ground, that's a counterproduc-tive exercise, squandering precious moisture. But in the fertile wet clays of northern Europe, the mouldboard plough was vastly superior, improving drainage and killing deep-rooted weeds, turning them from competition into compost.

The development of the mouldboard plough turned Europe's natural endowment of fertile land on its head. People who lived in northern Europe had long endured difficult farming conditions, but now it was the north, not the south, that enjoyed the best and most productive land. Starting about a thousand years ago, thanks to this new plough-based prosperity, cities of northern Europe emerged and started to flourish. And they flourished with a different social structure from cities around the Mediterranean. The dry-soil scratch plough needed only two animals to pull it, and it worked best with a criss-cross ploughing in simple, square fields. All this had made farming an individualistic practice: a farmer could live alone with his plough, oxen and land. But the wet-clay mouldboard plough required a team of eight oxen – or better, horses – and who had that sort of wealth? It was most efficient in long, thin strips often a step or two away from someone else's long, thin strips. As a result, farming became more of a community practice: people had to share the plough and draft animals, and resolve disagreements. They gathered together in villages. The mouldboard plough helped usher in the manorial system in northern Europe.

The plough also reshaped family life. It was heavy equipment, so ploughing was seen as men's work. But wheat and rice needed more preparation than nuts and berries, so women increasingly found themselves at home preparing food. There's a study of Syrian skeletons from nine thousand years ago which finds evidence that women were developing arthritis in their knees and feet, apparently from kneeling, twisting and grinding grain. And since women no longer had to carry toddlers around while foraging, they had more frequent pregnancies.

The plough-driven shift from foraging to farming may even have changed sexual politics. If you have land, that is an asset you can hand down to your children. And if you're a man, that means you might become increasingly concerned about whether they really are your children – after all, your wife is spending all her time at home while you are in the fields. Is she really doing nothing but grinding grain? So one theory – speculative but intriguing – is that the plough intensified men's policing of women's sexual activity. If that was an effect of the plough, it's been slow to fade.

The plough, then, did much more than increase crop yields. It changed everything, leading some to ask whether inventing the plough was entirely a good idea. Not that it didn't work – it worked brilliantly – but because along with providing the underpinnings of civilisation, it seems to have enabled the rise of misogyny and tyranny. Archaeological evidence also suggests that the early farmers had far worse health than their immediate hunter-gatherer forebears. With their diets of rice and grain, our ancestors were starved of vitamins, iron and protein. As societies switched from foraging to agriculture ten thousand years ago, the average height of both men and women shrank by around 6 inches (15cm), and there's ample evidence of parasites, disease and childhood malnutrition. Jared Diamond, author of *Guns, Germs and Steel*, called the adoption of agriculture 'the worst mistake in the history of the human race'.

You may wonder why, then, agriculture spread so quickly. We've already seen the answer: the food surplus enabled larger populations, and societies with specialists – builders, priests and craftsmen, but also specialist soldiers. Armies – even of stunted soldiers – will have been sufficiently powerful to drive the remaining hunter-gatherer tribes off all but the most marginal land. Even there, today's few remaining nomadic tribes still have a relatively healthy diet, with a rich variety of nuts, berries and animals. One Kalahari bushman was asked why his tribe hadn't copied its neighbours and picked up the plough. He replied, 'Why should we, when there are so many mongongo nuts in the world?'

So here you are, one of the few survivors of the end of civilisation. Will you reinvent the plough, and start the whole thing over again? Or should we be content with our mongongo nuts?

Introduction

Kalahari bushmen may not want to take up the plough, but modern civilisation still offers them some other potentially lucrative opportunities: a mere 100ml of cold-pressed mongongo nut oil currently retails for £25.38 on evitamins.com, courtesy of the Shea Terra Organics company. Apparently, it's very good for your hair.

Mongongo nut oil thus counts as one of the approximately ten billion distinct products and services currently offered in the world's major economic centres. The global economic system that delivers these products and services is vast and impossibly complex. It links almost every one of the planet's 7.5 billion people. It delivers astonishing luxury to hundreds of millions. It also leaves hundreds of millions behind, puts tremendous strains on the planet's ecosystem, and – as the financial meltdown of 2007–8 reminded us – has an alarming habit of spinning into the occasional crisis. Nobody is in charge of it. Indeed, no individual could ever hope to understand more than a fraction of what's going on.

How can we get our heads around this bewildering system on which our lives depend?

Another one of those ten billion products – this book – is an attempt to answer that question. Have a closer look at it.

(If you're listening to an audiobook or reading on a tablet, you'll have to summon up a memory of what a paper book feels like.) Just run your fingers over the surface of the paper. Isn't it remarkable? It's flexible, so that it can be bound into a book, the pages turning easily without an elaborate hinge. It's strong, so that it can be made in slim sheets. Just as important, it's cheap enough for many uses that will be more short-lived than this book. Cheap enough to use as a wrapping material, cheap enough to make newspapers that will be out of date within hours, cheap enough to use to wipe . . . well, to wipe anything you might want.

Paper is an amazing material, despite being throwaway stuff. In fact, paper is an amazing material in part *because* it's throwaway stuff. But there's more to a physical copy of this book than paper.

If you look at the back cover, you'll see a barcode, possibly more than one. The barcode is a way to write a number so that a computer can read it easily, and the barcode on the back of this book distinguishes it from every other book that has ever been written. Other barcodes distinguish Coca-Cola from industrial bleach, an umbrella from a portable hard drive. These barcodes are more than just a convenience at the checkout. The development of the barcode has reshaped the world economy, changing where products are made and where we're able to buy them. Yet the barcode itself is often overlooked.

Near the front of the book there's a copyright notice. It tells you that while this book belongs to you, the *words* in the book belong to *me*. What does that even mean? It's the result of a meta-invention, an invention *about* inventions – a concept called 'intellectual property'. Intellectual property has profoundly shaped who makes money in the modern world.

Yet there's an even more fundamental invention on display:

writing itself. The ability to write down our ideas, memories and stories underpins our entire civilisation. But we're now coming to realise that writing itself was invented for an economic purpose, to help coordinate and plan the comings and goings of an increasingly sophisticated economy.

Each of these inventions tells us a story, not just about human ingenuity, but about the invisible systems that surround us: of global supply chains, of ubiquitous information, of money and ideas and, yes, even of the sewage pipe that carries away the toilet paper we flush out of sight.

This book shines a spotlight on the fascinating details of the ways our world economy works by picking out fifty specific inventions – including paper, the barcode, intellectual property and writing itself. In each case, we'll find out what happens when we zoom in closely to examine an invention, or pull back to notice the unexpected connections. Along the way, we'll discover the answer to some surprising questions. For instance:

- What's the connection between Elton John and the promise of the paperless office?
- Which American discovery was banned in Japan for four decades, and how did that damage the careers of Japanese women?
- Why did police officers believe that they might have to execute a London murderer twice in 1803 – and what does that have to do with portable electronics?
- How did a monetary innovation destroy the Houses of Parliament?
- Which product was launched in 1976, flopped immediately, yet was lauded by the Nobel laureate economist Paul Samuelson alongside wine, the alphabet and the wheel?

- What does Federal Reserve chair Janet Yellen have in common with the great Mongolian-Chinese emperor Kublai Khan?

Some of these fifty inventions, such as the plough, are absurdly simple. Others, such as the clock, have become astonishingly sophisticated. Some of them are stodgily solid, like concrete. Others, such as the limited liability company, are abstract inventions that you cannot touch at all. Some, like the iPhone, have been insanely profitable. Others, like the diesel engine, were initially commercial disasters. But all of them have a story to tell that teaches us something about how our world works and that helps us notice some of the everyday miracles that surround us, often in the most ordinary-seeming objects. Some of those stories are of vast and impersonal economic forces; others are tales of human brilliance or human tragedy.

This book isn't an attempt to identify the fifty most economically significant inventions. It's not a book-length listicle, with a countdown to the most important invention of all. Indeed, some that would be no-brainers on any such list haven't made the cut: the printing press, the spinning jenny, the steam engine, the aeroplane and the computer.

What justifies such omissions? Simply that there are other stories to tell. For example, the attempt to develop a 'death ray' that led, instead, to radar, the invention that helps keep air travel safe. Or the invention that came to Germany shortly before Gutenberg invented the printing press, and without which printing would be technically feasible but economic suicide. (You've guessed it: paper.)

And I don't want to throw shade on the computer, I want to shed light. But that means looking instead at a cluster of inventions that turned computers into the remarkable multi-purpose tools they are today – Grace Hopper's compiler, which made communication between humans and computers much easier; public key cryptography, which keeps e-commerce secure; and the Google search algorithm, which makes the World Wide Web intelligible.

As I researched these stories, I realised that some themes emerged over and over again. The plough illustrates many of them: for example, the way new ideas often shift the balance of economic power, creating both winners and losers; how changes to the economy can have unexpected effects on the way we live, such as changing relationships between men and women; and how an invention like the plough opens up the possibility for further inventions such as writing, property rights, chemical fertiliser and much more.

So I've interspersed the stories with interludes to reflect on these common themes. And by the end of the book, we'll be able to draw these lessons together and ask how we should think about innovation today. What are the best ways to encourage new ideas? And how can we think clearly about what the effects of those ideas might be, and act with foresight to maximise the good effects and mitigate the bad ones?

It's all too easy to have a crude view of inventions – to see them simply as solutions to problems. Inventions cure cancer. Inventions get us to our holiday destination more quickly. Inventions are fun. Inventions make money. And of course it is true that inventions catch on because they do solve a problem that somebody, somewhere, wants to be solved. The plough caught on because it helped farmers to grow more food for less effort.

But we shouldn't fall into the trap of assuming that inventions are nothing but solutions. They're much more than that. Inventions shape our lives in unpredictable ways – and while they're solving a problem for someone, they're often creating a problem for someone else.

These fifty inventions that shaped our economy didn't do so by just producing more stuff, more cheaply. Each of them tugged on a complex web of economic connections. Sometimes they tangled us up, sometimes they sliced through old constraints, and sometimes they wove entirely new patterns.

1

WINNERS AND LOSERS

There's a word for those stubborn idiots who just don't understand the benefits of new technology: 'Luddite'. Economists – ever ready to adopt a bit of jargon – even speak of the 'Luddite fallacy', the dubious belief that technological progress creates mass unemployment. The original Luddites were weavers and textile workers who smashed mechanical looms in England two hundred years ago.

'Back then, some believed that technology would create unemployment. They were wrong,' comments Walter Isaacson, biographer of Albert Einstein, Ben Franklin and Steve Jobs. 'The industrial revolution made England richer and increased the total number of people in work, including in the fabric and clothing industries.'

Indeed it did. But to dismiss the Luddites as backward fools would be unfair. The Luddites didn't smash machine looms because they wrongly feared that machines would make England poorer. They smashed the looms because they rightly feared that machines would make *them* poorer. They were skilled workers who knew that the machine looms would devalue their skills. They understood perfectly well the implications of the technology they faced and they were right to dread it.

The Luddite predicament is not uncommon. New technologies almost always create new winners and losers. Even a better mousetrap is bad news for the manufacturers of traditional mousetraps. It is hardly good news for the mice, either.

And the process by which the playing field changes shape is not always straightforward. The Luddites weren't worried about being replaced by machines: they were worried about being replaced by the cheaper, less-skilled workers whom the machines would empower.

So whenever a new technology emerges, it's worth trying to ask who will win and who will lose out as a result. The answer can often surprise us.

2

The Gramophone

Who's the best paid solo singer in the world? In 2015, according to *Forbes*, it was probably Elton John. He reportedly made a hundred million dollars. U2 made twice as much as that, apparently, but there are four of them. There's only one Elton John.

Two hundred and fifteen years ago, the answer to the same question would have been: the best paid singer in the world is Mrs Billington. Elizabeth Billington was, some say, the greatest English soprano who ever lived. Sir Joshua Reynolds, the first president of the Royal Academy of Arts, once painted her, depicting her standing with a book of music in her hands and her curls partly pinned up and partly floating free, listening to a choir of angels singing. The composer Joseph Haydn thought the portrait was an injustice: the angels, said Haydn, should have been listening to Mrs Billington singing.

Elizabeth Billington was also something of a sensation off the stage. A scurrilous biography of her sold out in less than a day. The book contained what were purportedly copies of intimate letters about her famous lovers – including, they say, the Prince of Wales, the future King George IV. In a more dignified celebration of her fame, when she recovered from a six-week-long illness on her Italian tour, the Venice Opera House was illuminated for three days.

Such was Elizabeth Billington's fame – some would say notoriety – that she was the subject of a bidding war for her performances. The managers of what were then London's two leading opera houses, Covent Garden and Drury Lane, scrambled so desperately to secure her that she ended up singing at both venues, alternating between the two, and pulling in at least £10,000 in the 1801 season. It was, even for her, a remarkable sum, much noted at the time. But in today's terms, it's a mere £687,000, or about a million dollars – 1 per cent of Elton John's earnings.

What explains the difference? Why is Elton John worth a hundred Elizabeth Billingtons?

Nearly sixty years after Elizabeth Billington's death, the great economist Alfred Marshall analysed the impact of the electric telegraph. It then connected America, Britain, India and even Australia. Thanks to such modern communications, he wrote, 'men, who have once attained a commanding position, are enabled to apply their constructive or speculative genius to undertakings vaster, and extending over a wide area, than ever before.' The world's top industrialists were getting richer, faster. The gap between themselves and less outstanding entrepreneurs was growing.

But not every profession's best and brightest could gain in the same way, Marshall said. Looking for a contrast, he chose the performing arts. The 'number of persons who can be reached by a human voice', he observed, 'is strictly limited' – and so, in consequence, was the earning power of the vocalists.

Two years after Alfred Marshall wrote those words – in 1877, on Christmas Eve – Thomas Edison applied for a patent for the phonograph. It was the first machine to be able to both record and reproduce the sound of a human voice.

Nobody quite seemed to know what to do with the technology at first. A French publisher named Édouard-Léon Scott de Martinville had already developed something called the phonoautograph, a device intended to provide a visual record of the sound of a human voice – a little like a seismograph records an earthquake. But it does not seem to have occurred to Monsieur Martinville that one might try to convert the recording back into sound again.

Soon enough, the application of the new technology became clear: you could record the best singers in the world, and sell the recordings. At first making a recording was a bit like making carbon copies on a typewriter: a single performance could be captured only on three or four phonographs at once. In the 1890s, there was great demand to hear a song by the African-American singer George W. Johnson; to meet that demand he reportedly spent day after day singing the same song till his voice gave out – singing the song fifty times a day would produce a mere 200 records. When Emile Berliner introduced recordings on a disc, rather than Edison's cylinder, this opened the way to mass production. Then came radio and film. Performers like Charlie Chaplin could reach a global market just as easily as the men of industry whom Alfred Marshall had described.

For the Charlie Chaplins and Elton Johns of the world, new technologies meant wider fame and more money. But for the journeymen singers, it was a disaster. In Elizabeth Billington's day, many half-decent singers made a living performing live in music halls. Mrs Billington, after all, couldn't be everywhere. But when you can listen at home to the best performers in the world, why pay to hear a merely competent tribute act in person?

Thomas Edison's phonograph led the way towards a winner-take-all dynamic in the performing industry. The very best performers went from earning like Mrs Billington to earning like Elton John. Meanwhile, the only-slightly-less good went from making a comfortable living to struggling to pay their bills: small gaps in quality became vast gaps in money. In 1981, an economist called Sherwin Rosen called this phenomenon 'the Superstar economy'. Imagine, he said, the fortune that Mrs Billington might have made if there'd been phonographs in 1801.

Technological innovations have created superstar economics in other sectors, too. Satellite television, for example, has been to footballers what the gramophone was to musicians, or the telegraph to nineteenth-century industrialists. If you had been the world's best footballer a few decades ago, no more than a stadium-full of fans could have seen you play every week. Now, your every move will be watched by hundreds of millions on every continent. Part of the story is that football can be broadcast at all. But just as important was the growth in the number of television channels. When good football leagues became scarcer than broadcasters, the bidding war between those broadcasters became frenetic.

And as the market size for football expanded, so has the gap in pay between the very best and the merely very good. As recently as the 1980s, footballers in English football's top tier used to earn twice as much as those in the third tier, playing for – say – the fiftieth best team in the country. Now average wages in the Premier League are twenty-five times those earned by the players two divisions below.

Technological shifts can dramatically change who gets what, and they are wrenching because they can be so sudden – and because the people concerned have the same skills as ever, but suddenly have very different earning power.

Nor is it easy to know how to respond: when inequality is caused by a change in the tax code, by corporate collusion, or by governments favouring special interests, at least you have an enemy. But we can hardly ban Google and Facebook just to protect the livelihoods of newspaper reporters.

Throughout the twentieth century, new innovations – the cassette, the CD, the DVD – maintained the economic model created by the gramophone. But at the end of the century came the MP3 format and fast internet connections. Suddenly, you didn't have to spend twenty bucks on a plastic disc to hear your favourite music – you could find it online, free. In 2002, David Bowie warned his fellow musicians that they were facing a very different future: 'Music itself is going to become like running water or electricity,' he said. 'You'd better be prepared for doing a lot of touring because that's really the only unique situation that's going to be left.'

Bowie seems to have been right. Artists have stopped using concert tickets as a way to sell albums, and started using albums as a way to sell concert tickets. But we haven't returned to the days of Mrs Billington: amplification, stadium rock, global tours and endorsement deals mean that the most admired musicians can still profit from a vast audience. Inequality remains alive and well – the top 1 per cent of artists take more than five times more money from concerts than the bottom 95 per cent put together. The gramophone may be passé, but the ability of technological change to alter who wins and who loses is always with us.

3

Barbed Wire

Late in 1876, so the story goes, a young man named John Warne Gates built a wire-fence pen in the military plaza in the middle of San Antonio, Texas. He rounded up some of the toughest and wildest longhorns in all of Texas – or that's how he described them. Others say that the cattle were a docile bunch. And there are those who wonder whether this particular story is true at all. But never mind.

Gates – a man who later won the nickname 'Bet-A-Million Gates' – began to take bets with onlookers as to whether these powerful, ornery longhorns could break through the fragile-seeming wire. They couldn't.

Even when Gates's sidekick, a Mexican cowboy, charged at the cattle howling Spanish curses and waving a burning brand in each hand, the wire held. Bet-A-Million Gates wasn't so worried about winning his wagers. He had a bigger game to play: he was selling a new kind of fence, and the orders soon came rolling in.

The advertisements of the time touted this fence as 'The Greatest Discovery Of The Age', patented by J.F. Glidden of De Kalb, Illinois. John Warne Gates described it more poetically: 'Lighter than air, stronger than whiskey, cheaper than dust.' We simply call it barbed wire.

To claim that barbed wire is the greatest discovery of the age might seem hyperbolic, even making allowances for the fact that the advertisers didn't know that Alexander Graham Bell was just about to be awarded a patent for the telephone. But while modern minds naturally think of the telephone as transformative, barbed wire wreaked huge changes on the American west, and much more quickly.

Joseph Glidden's design for barbed wire wasn't the first, but it was the best. It is recognisably modern: it is the same as the barbed wire you can see around farmland today. The wicked barb is twisted around a strand of smooth wire, then a second strand of smooth wire is twisted together with the first to stop the barbs from sliding around. Farmers snapped it up.

There was a reason that American farmers were so hungry for barbed wire. A few years earlier, in 1862, President Abraham Lincoln had signed the Homestead Act. The act specified that any honest citizen – including women and freed slaves – could lay claim to up to 160 acres of land in America's western territories. All they had to do was build a home there and work the land for five years. The idea was that the Homestead Act would improve both the land and the lot of American citizens, creating free and virtuous hardworking landowners with a strong stake in the future of the nation.

It sounds simple. But the prairie was a vast and uncharted expanse of tall, tough grasses – a land suitable for nomads, not settlers. It had long been the territory of the Native Americans. After Europeans arrived and pushed west, the cowboys roamed free, herding cattle over the boundless plains.

But settlers needed fences, not least to keep those

free-roaming cattle from trampling their crops. And there wasn't a lot of wood – certainly none to spare for fencing in mile after mile of what was often called 'The American Desert'. Farmers tried cultivating thornbush hedges, but they were slow-growing and inflexible. Smooth wire fences didn't work either – the cattle simply pushed through them.

The lack of fencing was much lamented. The US Department of Agriculture conducted a study in 1870 and concluded that until one of those technologies worked, it would be impossible to settle the American west. The American west, in turn, seethed with potential solutions: at the time, it was the source of more proposals for new fencing technologies than the rest of the world put together. And the idea that emerged in triumph from this intellectual ferment? Barbed wire.

Barbed wire changed what the Homestead Act could not. Until barbed wire was developed, the prairie was an unbounded space, more like an ocean than a stretch of arable land. Private ownership of land wasn't common because it wasn't feasible.

So barbed wire spread because it solved one of the biggest problems that the settlers faced. But it also sparked ferocious disagreements. And it's not hard to see why. The homesteading farmers were trying to stake out their property – property that had once been the territory of various Native American tribes. And twenty-five years after the Homestead Act came the Dawes Act, which forcibly assigned land to Native American families and gave the rest to white farmers. Olivier Razac, the author of a book on barbed wire, comments that as well as freeing up land for settler cultivation, the Dawes Act 'helped destroy the foundations of Indian society'. No wonder those tribes called barbed wire 'The Devil's Rope'.

The old-time cowboys also lived on the principle that

cattle could graze freely across the plains – this was the law of the open range. The cowboys hated the wire: cattle would get nasty wounds and infections. When the blizzards came, the cattle would try to head south; sometimes they got stuck against the wire and died in their thousands.

Other cowmen adopted barbed wire, using it to fence off private ranches. And while the attraction of barbed wire was that it could enforce legal boundaries, many of the fences were illegal, too – attempts to commandeer common land for private purposes.

When the barbed wire fences started to go up across the west, fights started to break out. In the 'fence-cutting wars', masked gangs with names such as the Blue Devils and the Javelinas cut the wires and left death threats warning the fence-owner not to rebuild. There were shoot-outs, even a few deaths. Eventually the authorities clamped down. The fence-cutting wars ended; the barbed wire remained. There were winners, and there were losers.

'It makes me sick,' said one trail driver in 1883, 'when I think of onions and Irish potatoes growing where mustang ponies should be exercising and where four-year-old steers should be getting ripe for market.' And if the cowboys were outraged, the Native Americans were suffering far worse.

These ferocious arguments on the frontier reflected an old philosophical debate. The English seventeenth-century philosopher John Locke – a great influence on the founding fathers of the United States – puzzled over the problem of how anybody might legally come to own land. Once upon a time, nobody owned anything: land was a gift of nature or of God. But Locke's world was full of privately owned land, whether the owner was the King himself or a simple yeoman. How had nature's bounty become privately owned? Was that inevitably the result of a guy with a bunch of goons grabbing whatever he could? If so, all civilisation was built on violent theft. That wasn't a welcome conclusion to Locke – or to his wealthy patrons.

Locke argued that we all own our own labour. And if you mix your labour with the land that nature provides – for example, by ploughing the soil – then you've blended something you definitely own with something that nobody owns. By working the land, he said, you've come to own it.

This wasn't a purely theoretical argument. Locke was actively engaged in the debate over Europe's colonisation of America. Political scientist Barbara Arneil, an expert on Locke, writes that 'the question, "How was private property created by the first men?" is for Locke the same question as, "Who has just title to appropriate the lands of America now?".' And to make his argument, he also had to make the claim that the land was abundant and unclaimed – that is, that because the indigenous tribes hadn't 'improved' the land, they had no right to it.

Not every European philosopher bought this line of argument. Jean-Jacques Rousseau, the eighteenth-century philosopher, protested the evils of enclosure. In his 'Discourse on Inequality' he lamented 'The first man who, having enclosed a piece of ground, bethought himself of saying "This is mine" and found people simple enough to believe him.' This man, said Rousseau, 'was the real founder of civil society'.

Rousseau did not intend that as a compliment. But complimentary or not, it's true that modern economies are built on private property – on the legal fact that most things have an owner, usually a person or a corporation. Modern economies are also built on the idea that private property is a good thing, because private property gives people an incentive to invest in and improve what they own – whether that's a patch of land in the American mid-west, or an apartment in Kolkata, or even a piece of intellectual property such as the rights to Mickey Mouse. It's a powerful argument – and it was ruthlessly deployed by those who wanted to argue that Native Americans didn't really have a right to their own territory, because they weren't actively developing it.

But legal facts are abstract. To get the benefits of owning something, you also have to be able to assert control over it.* Barbed wire is still widely used to fence off land across the world. And in many other spheres of the economy, the battle to own in practice what you own in theory continues to rage.

Musicians may have copyright in their music, but – as David Bowie kindly explained to us – copyright is a weak defence against file-sharing software.

And nobody has invented virtual barbed wire that can fence off songs as effectively as physical barbed wire fenced off land – but it hasn't stopped people trying. The 'fence-cutting wars' of the digital economy are no less impassioned today than they were in the Wild West: digital rights campaigners battle the likes of Disney, Netflix and Google, while hackers and pirates make short work of the digital barbed wire. When it comes to protecting property in any economic system, the stakes are very high.

Small wonder that the barbed wire barons – Bet-A-Million Gates, Joseph Glidden and several others – became rich. The year that Glidden secured his barbed wire patent, 32 miles of wire were produced. Six years later, in 1880, the factory in DeKalb turned out 263,000 miles of wire, enough to circle the world ten times over.

---

* Until barbed wire was developed, settlers in the American west had legal rights over their land, but no way of exerting practical control. Later in the book, we'll discuss the mirror image of that situation: countries where people have practical control over their homes and farms, but no legal rights.

4

Seller Feedback

In Shanghai, a driver logs into an online forum, looking for someone to pretend to want a ride. He finds a willing taker. He pretends to collect the customer and drop her at the airport; in fact, they never meet. Then he goes online and sends her money. The fee they'd agreed is about $1.60.

Or perhaps the driver goes a step further, making up not just the journey, but also the other person. He goes to the online marketplace Taobao, and buys a hacked smartphone. That enables him to create multiple fake identities; he uses one to arrange a ride with himself.

Why is he doing this? Because he's willing to run the risk of being caught – and because someone's willing to pay him to give people rides in his car. Investors have run up billions of dollars of losses – in China, and elsewhere – paying people to share car journeys. Naturally, they're trying to stamp out the imaginary journeys, but subsidising genuine rides? They're convinced that's a smart idea.

This all seems bizarre – even perverse. But everyone involved is rationally pursuing economic incentives. To see what's going on, we have to understand a phenomenon that's spawned many buzzwords: 'crowd-based capitalism', 'collaborative consumption', 'the sharing economy' and 'the trust economy'.

Here's the basic idea. Suppose I'm about to drive myself from downtown Shanghai to the airport. I occupy only one seat in my car. Now suppose that you live a block away, and you also need to catch a flight. Why don't I give you a lift? You could pay me a modest sum, less than you'd pay for some other mode of transport. You're better off. So am I – after all, I was driving to the airport anyway.

There are two big reasons why this might not happen. The first, and most obvious, is if neither of us knows the other exists. Until recently, the only way you could advertise your desire for a lift would be to stand at an intersection, holding up a sign saying 'airport'. It's not very practical – especially since the plane won't wait.

Other transactions are even more niche. Say I'm working at home, and my dog is nuzzling my leg, his leash in his mouth, desperate for a walk. But I'm behind on a deadline and can't spare the time. You, meanwhile, live nearby. You like dogs, and walking, and have a free hour. You'd love to earn a few bucks by walking my dog, and I'd love to pay you. How do we find each other? We don't – unless we have some kind of online platform, something like TaskRabbit, or Rover.

This function of matching people who have coincidental wants is among the most powerful ways the internet is reshaping the economy. Traditional markets work perfectly well for some goods and services, but they're less useful when the goods and services are urgent or obscure.

Consider the plight of Mark Fraser. It was 1995. Mark Fraser gave lots of presentations, and he *really* wanted a laser pointer – they were new, and cool, but also forbiddingly expensive. Fraser, however, was an electronics geek. He was confident that if he could get his hands on a broken laser pointer, he could fix it up. But where on earth would he find a broken laser pointer? The answer, now, is obvious – try Taobao, or eBay, or some other online marketplace. Back then, eBay had only just started. Its very first sale: Mark Fraser bought a broken laser pointer.

Mark Fraser was taking a bit of a risk. He didn't know the seller; he simply had to trust that they wouldn't pocket his $14.83 and disappear. For other transactions, the stakes are higher. That's the second reason I might not give you a lift to Shanghai airport. I see you at the intersection, holding your sign – but I've no idea who you are. Perhaps you're intending to attack me and steal my car? You might doubt my motives, too – perhaps I'm a serial killer.

That's not a completely ridiculous concern: hitch-hiking was a popular pursuit a few decades ago, but after some sensationally publicised murders, it fell out of fashion.

Trust is an essential component of markets – so essential that we often don't even notice it, like a fish doesn't notice water. In developed economies, enablers of trust are everywhere: brands, money-back guarantees, and of course repeat transactions with a seller who can be easily located.

But the new sharing economy lacks those enablers. Why should we get into a stranger's car – or buy a stranger's laser pointer? In 1997, eBay introduced a feature that helped solve the problem: Seller Feedback. Jim Griffith was eBay's first customer service representative; at the time, he says, 'no-one had ever seen anything like [it]'. The idea of both parties rating each other after a transaction has now become ubiquitous. You buy something online – you rate the seller, the seller rates you. You use a ride-sharing service, like Uber – you rate the driver, the driver rates you. You stay in an Airbnb – you rate the host, the host rates you. Analysts like

Rachel Botsman reckon the 'reputation capital' we build on such websites will eventually become more important than credit scores. Possibly, but these systems aren't bulletproof. However, they achieve a crucial basic job – helping people to overcome natural caution.

A few positive reviews set our mind at ease about a stranger. Jim Griffith says of Seller Feedback, 'I'm not so sure [eBay] would have grown without it'. Online matching platforms would still exist, of course – eBay already did – but perhaps they'd be more like hitch-hiking today: a niche pursuit for the unusually adventurous, not a mainstream activity that's transforming whole sectors of the economy.

Platforms like Uber and Airbnb, eBay and TaskRabbit create real value. They tap into capacity that would have gone to waste: a spare room, a spare hour, a spare car seat. They help cities be flexible when there are peaks in demand: I might let out a room only occasionally, when some big event means the price is high.

But there are losers. For all the touchy-feeliness of the buzzwords – 'collaborative', 'sharing', 'trust' – these models aren't all about heartwarming stories of neighbours coming together to borrow each other's power drills. They can easily lead to cut-throat capitalism. Established hotels and taxi companies are aghast at competition from Airbnb and Uber. Is that just an incumbent trying to suppress competition? Or are they right when they complain that the new platforms are ignoring important regulations?

Many countries have rules to protect workers, like guaranteed hours or working conditions or a minimum wage. And many people on platforms like Uber aren't just monetising spare capacity, they're trying to make a living, without those protections of a formal job; perhaps because those very platforms competed them out of a job.

Some regulations protect customers, too – for example, from discrimination. Hotels can't legally refuse you a room if you're, say, a same-sex couple. But hosts on Airbnb can choose to turn down guests after seeing not just your feedback but your photos. Airbnb builds trust by bigging-up the personal connection, and that means showing people prominent pictures of who they're dealing with. It also enables people to act on their personal prejudices, consciously or otherwise. People from ethnic minorities have been proven to suffer as a result. How online matching platforms should be regulated is a dilemma causing lawmakers around the world to scratch their heads.

It matters because it's potentially huge business, especially in emerging markets where there isn't yet a culture of owning things like cars. And it's a business with network effects: the more people use a platform, the more attractive it becomes. That's why Uber and its rivals – Didi Chuxing in China, Grab in southeast Asia, Ola in India – have invested massively in subsidising rides and giving credits to new customers: they wanted to get big first.

And, naturally, some drivers have been tempted to defraud them. Remember how they did it? By using an online forum to find a willing fake customer, or an online marketplace to buy a hacked smartphone. Matching people with particular wants really is useful.

5

Google Search

'Dad, what happens when you die?'

'I don't know, son. Nobody knows for sure.'

'Well, why don't you ask Google?'

Evidently, it's possible for children to grow up with the impression that Google knows *everything*. Perhaps that's to be expected. 'Dad, how far is the Moon from the Earth?' 'What's the biggest fish in the world?' 'Do jetpacks really exist?' All efficiently answered with the tap of a touchscreen. No need to visit the library to consult the *Encyclopaedia Britannica*, the *Guinness Book of Records*, or – well, who knows how a pre-Google parent would have discovered the state of the art in jetpack technology. It wouldn't have been straightforward.

Google may not be clever enough to know if there's life after death, but the word 'google' does crop up in conversation more often than either 'clever' or 'death', according to researchers at the UK's Lancaster University. It took just two decades for Google to reach this cultural ubiquity, from its humble beginnings as a student project at Stanford University.

It's hard to remember just how bad search technology was before Google. In 1998, for example, if you typed 'cars' into

Lycos – then a leading search engine – you'd get a results page filled with porn websites. Why? Owners of porn websites inserted many mentions of popular search terms like 'cars', perhaps in tiny text, or in white on a white background. The Lycos algorithm saw many mentions of 'cars', and concluded that the page would be interesting to someone searching for 'cars'. It's a system that now seems almost laughably simplistic, and easy to game.

Larry Page and Sergey Brin were not, initially, interested in designing a better way to search. Their Stanford project had a more academic motivation. In academia, how often a published paper is cited is a measure of how much credibility it has; and if it's cited by papers which themselves are cited many times, that bestows even more credibility. Page and Brin realised that when you looked at a page on the nascent World Wide Web, you had no way of knowing which other pages linked to it. Web links are analogous to academic citations. If they could find a way to analyse all the links on the web, they could rank the credibility of each web page in any given subject.

To do this, Page and Brin first had to download the entire internet. This caused some consternation. It gobbled up nearly half of Stanford's bandwidth. Irate webmasters showered the university with complaints that Google's crawler was overloading their servers. An online art museum thought Stanford was trying to steal their content, and threatened to sue. But as Page and Brin refined their algorithm, it soon became clear that they had stumbled on a new and vastly better way to search the web. Put simply, porn websites with tiny text saying 'cars cars cars' don't get many links from other websites that discuss cars. If you searched Google for 'cars', its analysis of the web's network of links would be likely to yield results about ... cars.

With such an obviously useful product, Page and Brin attracted investors, and Google went from student project to private company. It's now among the world's biggest, bringing in profits by the tens of billions of dollars. But for the first few years, Page and Brin burned through money without much idea about how they'd ever make it back. They weren't alone. This was the time of the dotcom boom and bust – shares in loss-making internet companies traded at absurd prices, based purely on hope that eventually they'd figure out viable business models.

It was 2001 when Google found theirs, and in retrospect it seems obvious: pay-per-click advertising. Advertisers tell Google how much they'll pay if someone clicks through to their website, having searched for terms they specify. Google displays ads from the highest bidders alongside its 'organic' search results. From an advertiser's perspective, the appeal is clear: you pay only when you reach people who have just demonstrated an interest in your offering. (Try Googling 'what happens when you die': there's an advertiser willing to pay Google for your click-through – the Mormons.) That's much more efficient than paying to advertise in a newspaper: even if its readership matches your target demographic, inevitably most people who see your advert won't be interested in what you're selling. No wonder newspaper advertising revenue has fallen off a cliff.

The media's scramble for new business models is one obvious economic impact of Google search. But the invention of functional search technology has created value in many ways. A few years ago, consultants at McKinsey tried to list the most important.

There are time savings. Studies suggest that Googling is about three times as quick as finding information in a library, and that's before you count the time spent travelling to the library. Likewise, finding a business online is about three times faster than using a traditional, printed directory like the Yellow Pages. McKinsey put the productivity gains into the hundreds of billions of dollars.

Another benefit is price transparency – that's economist jargon for being able to stand in a store, take out your phone, Google a product you're thinking of buying to see if it's available more cheaply elsewhere, and use that knowledge to haggle. Annoying for the store, helpful for the customer.

Then there are 'long tail' effects. In physical stores, space is at a premium. Online stores can offer more variety – but only when the search engines are good enough to enable customers to find what they need. Online shopping with a search function that works means customers with specific desires are likelier to find exactly what they want, rather than having to settle for the closest thing available locally. And it means entrepreneurs can launch niche products, more confident that they'll find a market.

This all sounds like excellent news for consumers and businesses. But there are problems.

One problem is those advertisements. Typically they function as one might expect – if you Google 'craft beer' then you'll get adverts for craft beer – but certain kinds of search attract fly-by-night companies bidding handsomely for a click-through from people who are in a bind. Google 'locksmith near me', for example, and your top results might include plausible-looking outfits that quote a low price for getting you back into your house – which becomes a much higher price when their representative arrives on your doorstep and claims to discover an unanticipated complication. Similar distressed-search adverts exist for people who've lost their wallet in the back of a New York taxicab, or who need to rebook a flight at short notice. In a panic, they don't notice that they're not getting quite what they expect at the end of the search result. Some of these companies are outright fraudulent; others cleverly tiptoe up to the line without crossing it. Nor is it clear how hard Google is trying to stamp out this sort of thing.

Perhaps the bigger issue is that this seems to be Google's responsibility alone, because the company dominates the search market. Google handles close to 90 per cent of searches worldwide; many businesses rely on ranking highly in its organic search results. And Google constantly tweaks the algorithm that decides them. Google gives general advice about how to do well, but it isn't transparent about how it ranks results. Indeed, it can't be: the more Google reveals, the easier it is for the scammers. We'd be back to searching for cars and getting porn.

You don't have to look far online – starting with Google, naturally – to find business owners and search strategy consultants gnashing their teeth over the company's power to make or break them. If Google thinks you're employing tactics it considers unacceptable, it'll downgrade you. One blogger complains that Google is 'judge, jury and executioner . . . you get penalized on suspicion of breaking the rules [and] you don't even know what the rules are, you can only guess'. Trying to figure out how to please Google's algorithm is rather like trying to appease an omnipotent, capricious and ultimately unknowable deity.

You may say this is no problem. As long as Google's top results are useful to searchers, it's tough luck on those who rank lower – and if those results stop being useful, then some other pair of students at Stanford will spot the gap in the market and dream up a better way. Right? Maybe – or maybe not. Search was a competitive business in the late 1990s. But now, it may be a natural monopoly – in other words, an industry that's extremely hard for a new entrant to succeed in.

The reason? Among the best ways to improve the usefulness of search results is to analyse which links were ultimately clicked by people who previously performed the same search, as well as what the user has searched for before. Google has far more of that data than anyone else. That suggests it may continue to shape our access to knowledge for generations to come.

6

Passports

'What would we English say if we could not go from London to the Crystal Palace or from Manchester to Stockport without a passport or police officer at our heels? Depend upon it, we are not half enough grateful to God for our national privileges.'

Those are the musings of an English publisher named John Gadsby, travelling through Europe in the mid-nineteenth century. This was before the modern passport system, wearily familiar to anyone who's ever crossed a national border: you stand in a queue; you proffer your standardised booklet to a uniformed official who glances at your face to check that it resembles the image of your younger, slimmer self (that haircut: what were you thinking?). Perhaps the official quizzes you about your journey, while the computer checks your name against a terrorist watchlist.

For most of history, passports were neither so ubiquitous nor so routinely used. They were, essentially, a threat: a letter from some powerful person requesting anyone the traveller met to let them pass unmolested – or else. The concept of passport as protection goes back to biblical times. And protection was a privilege, not a right: English gentlemen like Gadsby who wanted a passport before venturing across the Channel once needed to unearth some personal social link to the relevant government minister.

As Gadsby discovered, the more zealously bureaucratic of Continental nations had realised the passport's potential as a tool of social and economic control. Even a century earlier, the citizens of France had to show paperwork not only to leave the country, but to travel from town to town. While wealthy countries today secure their borders to keep unskilled workers out, municipal authorities historically used them to stop their skilled workers from leaving.

As the nineteenth century progressed, the railways and the steamboat made travel faster and cheaper. Passports were unpopular. France's Emperor Napoleon III shared Gadsby's admiration for the more relaxed British approach: he described passports as 'an oppressive invention ... an embarrassment and an obstacle to the peaceable citizen'. He abolished them in 1860. France was not alone. More and more countries either formally abandoned passport requirements or stopped bothering to enforce them, at least in peacetime. You could visit 1890s America without a passport, though it helped if you were white. In some South American countries, passport-free travel was in the constitution. In China and Japan, foreigners needed passports only to venture inland.

By the turn of the twentieth century, only a handful of countries were still insisting on passports to enter or leave. It seemed possible that passports might soon disappear altogether.

What would today's world look like if they had?

Early one morning in September 2015, Abdullah Kurdi, his wife and two young sons boarded a rubber dinghy on a beach in Bodrum, Turkey. They hoped to make it 4km across the Aegean Sea to the Greek island of Kos. But the sea became rough, and the dinghy capsized. Abdullah managed to cling to the boat, but his wife and children drowned.

The body of his youngest child, three-year-old Alan, washed up on a Turkish beach, where it was photographed by a Turkish agency journalist. The image of Alan Kurdi became an icon of the migrant crisis that had convulsed Europe all summer.

The Kurdis hadn't planned to stay in Greece. They hoped eventually to start a new life in Vancouver, where Abdullah's sister Teema worked as a hairdresser. There are easier ways to travel from Turkey to Canada than starting with a dinghy to Kos, and Abdullah had the money – the four thousand euros he paid a people-smuggler could instead have bought plane tickets for them all. At least, it could have done – if they hadn't needed the right passport.

Since the Syrian government denied citizenship to ethnic Kurds, the Kurdis had no passports. But even with Syrian passports, they couldn't have boarded a plane to Canada. If they'd had passports issued by Sweden or Slovakia, or Singapore or Samoa, they'd have had no problems.

It can seem like a natural fact of life that the name of the country on our passport determines where we can travel and work – legally, at least. But it's a relatively recent historical development, and, from a certain angle, it's odd. Your access to a passport is, broadly speaking, determined by where you were born and the identity of your parents. (Although if you've got $250,000, for example, you can buy one from St Kitts and Nevis.)

In most facets of life we want our governments and our societies to help overcome such accidents. Many countries take pride in banning employers from discriminating among workers based on characteristics we can't change: whether we're male or female, young or old, gay or straight, black or white. But when it comes to your citizenship, that's an accident of birth that we expect governments to preserve, not erase. And the passport is a crucial tool for ensuring that different people with different nationalities have access to very different opportunities.

There's no public clamour to judge people not by the colour of their passport, but by the content of their character. Less than three decades after the fall of the Berlin Wall, migrant controls are back in fashion. Donald Trump promises a wall along the US-Mexico border. The Schengen zone cracks under the pressure of the migrant crisis. Europe's leaders scramble to distinguish refugees from 'economic migrants', the assumption being that someone who isn't fleeing persecution – someone who merely wants a better job, a better life – should not be let in. Politically, the logic of restrictions on migration is increasingly hard to dispute.

Yet economic logic points in the opposite direction. In theory, whenever you allow factors of production to follow demand, output rises. In practice, all migration creates winners and losers, but research indicates that there are many more winners – in the wealthiest countries, by one estimate, five in six of the existing population are made better off by the arrival of immigrants.

So why does this not translate into popular support for open borders? There are practical and cultural reasons why migration can be badly managed: if public services aren't upgraded quickly enough to cope with new arrivals, or belief systems prove hard to reconcile. Also, the losses tend to be more visible than the gains. Suppose a group of Mexicans arrive in America, ready to pick fruit for lower wages than Americans are earning. The benefits – slightly cheaper fruit for everyone – are too widely spread and small to notice, while the costs – some Americans losing their jobs – produce vocal unhappiness. It should be possible to arrange taxes and public spending to compensate the losers. But it doesn't tend to work that way.

The economic logic of migration often seems more compelling when it doesn't involve crossing national borders. In 1980s Britain, with recession affecting some of the country's regions more than others, employment minister Norman Tebbit implied – or was widely taken to be implying – that the jobless should 'get on their bikes' to look for work. How much might global economic output rise if anyone could get on their bikes to work anywhere? Some economists have calculated it would double.

That suggests our world would now be much richer if passports had died out in the early twentieth century. There's one simple reason they didn't: the First World War intervened.

With security concerns trumping ease of travel, governments imposed strict new controls on movement – and they proved unwilling to relinquish their powers once peace returned. In 1920, the newly formed League of Nations called an 'International Conference on Passports, Customs Formalities and Through Tickets', and that effectively invented the passport as we know it. From 1921, the conference said, passports should be 15.5cm by 10.5cm; 32 pages; bound in cardboard; with a photo. The format has changed remarkably little since.

Like John Gadsby, anyone with the right colour passport can only count their blessings.

7

Robots

It's about the size and shape of an office photocopier. With a gentle whirring noise, it traverses the warehouse floor while its two arms raise or lower themselves on scissor lifts, ready for the next task. Each arm has a camera on its knuckle. The left arm eases a cardboard box forward on the shelf; the right arm reaches in and extracts a bottle.

Like many new robots, this one comes from Japan. The Hitachi corporation showcased it in 2015, with hopes to be selling it by 2020. It's not the only robot that can pick a bottle off a shelf – but it's as close as robots have yet come to performing this seemingly simple task as speedily and dextrously as a good old-fashioned human.

One day, robots like this might replace warehouse workers altogether. For now, humans and machines are running warehouses together: in Amazon's depots, the company's Kiva robots scurry around – not picking things off shelves, but carrying the shelves to humans for them to pick things off. By saving the time workers would otherwise spend trudging up and down aisles, Kiva robots can improve efficiency up to fourfold.

Robots and humans are working side-by-side in factories, too. Factories have had robots for decades, of course – since 1961, when General Motors installed the first Unimate, a one-armed robot resembling a small tank that was used for tasks like welding. But, until recently, they were strictly segregated from the human workers – partly to stop the humans coming to any harm, and partly to stop humans confusing the robots, whose working conditions had to be strictly controlled.

With some new robots, that's no longer necessary. A charming example by the name of Baxter can generally avoid bumping into humans, or falling over if humans bump into it. Baxter has cartoon eyes that help indicate to human co-workers where it is about to move. And if someone knocks a tool out of Baxter's hand, it won't dopily try to continue the job. Historically, industrial robots needed specialist programming; Baxter can learn new tasks from co-workers showing it what to do.

The world's robot population is expanding quickly – sales of industrial robots are growing around 13 per cent a year, which means the robot 'birth rate' is almost doubling every five years. For years, there's been a trend to 'offshore' manufacturing to emerging markets, where workers are cheaper; now, the trend is to 'reshore', and robots are part of that. Robots are doing more and more things. They're lettuce-pickers, bartenders and hospital porters. Still, let's face it: they're not yet doing as much as we'd once expected. In 1962, a year after the Unimate, the American cartoon *The Jetsons* imagined Rosie, a robot maid, doing all the household chores. Half a century on, where's Rosie? She's not coming any time soon, despite recent progress.

That progress is partly thanks to robot hardware – in particular, better and cheaper sensors. In human terms that's like improving a robot's eyes, the touch of its fingertips, or its inner ear – its sense of balance. But it's also thanks to software – in human terms, robots are getting better brains.

And it's about time: machine thinking is another area where early expectations were disappointed. Attempts to invent artificial intelligence are generally dated to 1956, and a summer workshop at Dartmouth College for scientists with a pioneering interest in 'machines that use language, form abstractions and concepts, solve kinds of problems now reserved for humans, and improve themselves'. At the time, machines with human-like intelligence were often predicted to be about twenty years away. Now, they're often predicted to be ... well, about twenty years away.

The futurist philosopher Nick Bostrom has a cynical take on this: twenty years is 'a sweet spot for prognosticators of radical change', he writes: nearer, and you'd expect to be seeing prototypes by now; further away, and it's not so attention-grabbing. Besides, says Bostrom, '[t]wenty years may also be close to the typical duration remaining of a forecaster's career, bounding the reputational risk of a bold prediction.'

It's only in the last few years that progress in artificial intelligence has really started to accelerate. Specifically, in what's known as *narrow AI* – algorithms that can do one thing very well, like playing Go, or filtering email spam, or recognising faces in your Facebook photos. Processors have got faster, data sets bigger and programmers better at writing algorithms that can learn to improve their own functioning, in ways that often remain opaque to their human creators.

That capacity for self-improvement causes some thinkers, like Bostrom, to worry what will happen if and when we create *artificial general intelligence* – a system that could apply itself to any problem, like humans can. Will it rapidly turn itself into a superintelligence? How would we keep it under control? That's not an imminent concern, at least; it's reckoned human-level artificial general intelligence is still about, ooh, twenty years away.

But narrow AI is already transforming the economy. For years, algorithms have been taking over white-collar drudgery in areas like bookkeeping and customer service. And more prestigious jobs are far from safe. IBM's Watson, which hit the headlines for beating human champions at the game show *Jeopardy*, is already better than doctors at diagnosing lung cancer. Software is getting to be as good as experienced lawyers at predicting what lines of argument are most likely to win a case. Robo-advisers are dispensing investment advice. And algorithms are routinely churning out news reports on subjects like the financial markets and sports – although, luckily for me, it seems they can't yet write books on economics.

Some economists reckon robots and AI explain a curious economic trend. Erik Brynjolfsson and Andrew McAfee argue there's been a 'great decoupling' between jobs and productivity – that's a measure of how efficiently an economy takes inputs, like people and capital, and turns them into useful stuff. Historically, as you'd expect, better productivity meant more jobs and higher wages. But Brynjolfsson and McAfee argue that's no longer the case in the United States. Since the turn of the century, US productivity has been improving, but jobs and wages haven't kept pace. Some economists worry that we're experiencing 'secular stagnation' – where there's not enough demand to spur economies into growing, even with interest rates down to zero or below.

The idea that technology can destroy or degrade some jobs isn't new – that's why the Luddites smashed machine looms two hundred years ago. But as we've seen, 'Luddite' has become a term of mockery because technology has always, eventually, created new jobs to replace the ones it destroyed. Those new jobs have tended to be better jobs – at least on average. But they haven't always been better, either for the workers or for society as a whole. An example: one dubious benefit of cash machines is that they freed up bank tellers to cross-sell dodgy financial products. What happens this time remains debatable: it's at least conceivable that some of the jobs humans will be left doing will actually be worse.

That's because technology seems to be making more progress at thinking than doing: robots' brains are improving faster than their bodies. Martin Ford, the author of *Rise of the Robots*, points out that robots can land aeroplanes and trade shares on Wall Street, but they still can't clean toilets.

So perhaps, for a glimpse of the future, we should look not to Rosie the Robot but to another device now being used in warehouses: the Jennifer Unit. It's a headset that tells human workers what to do, down to the smallest detail: so, if you have to pick nineteen identical items from a shelf, it'll tell you to pick five, then five, then five, then four … that leads to fewer errors than saying 'pick 19'. If robots beat humans at thinking, but humans beat robots at picking things off shelves, why not control a human body with a robot brain? It may not be a fulfilling career choice, but you can't deny the logic.

8

The Welfare State

Women in politics are sometimes accused of consciously exploiting their femininity to get ahead in a male-dominated world. Frances Perkins did that, but in an unusual way: she tried to remind men of their mothers. She dressed in a plain, three-cornered hat, and she refined the way she acted, based on careful observation of what seemed to be most effective in persuading men to accept her ideas.

Perhaps it's no coincidence that those ideas could reasonably be described as maternal – or, at least, parental. Any parent wants to shield their children from serious harm, and Perkins believed governments should do the same for their citizens. She became President Franklin D. Roosevelt's Secretary of Labor in 1933. The Great Depression was ravaging America: a third of workers were unemployed, and those in jobs saw their wages plunge. Perkins drove through the reforms that became known as the New Deal, including a minimum wage, benefits for the unemployed and pensions for the elderly.

Historians will tell you that it wasn't Frances Perkins who invented the welfare state. It was Otto von Bismarck,

Chancellor of the German Empire half a century earlier. But it was largely during Perkins's era that various welfare states took their recognisably modern shape across the developed world. Details differ, from place to place, measure to measure, and time to time. For some benefits, you have to have paid in to state-run insurance schemes; others are rights, based on residence or citizenship. Some benefits are universal – everyone gets them, regardless of income. Some are means-tested – you have to prove that you meet criteria of neediness.

But the same basic idea links every welfare state: that the ultimate responsibility for ensuring people don't starve on the street should lie not with family, or charity, or private insurers, but with government.

This idea is not without its enemies. It is possible, after all, to mother too much. Every parent instinctively knows that there's a balance: protect, but don't mollycoddle; nurture resilience, not dependence. And if overprotective parenting stunts personal growth, might too-generous welfare states stunt economic growth?

It's a plausible worry. Imagine a single mother with two children. She might qualify for various payouts: housing benefit, child benefit and unemployment benefit. Could she accumulate more from the welfare system than she could get by working at the minimum wage? In 2013, in no fewer than nine European countries, the answer to that question was 'yes'. Now it might still be attractive for this hypothetical woman to both work and draw benefits, but in three countries – Austria, Croatia and Denmark – her marginal tax rate was nearly a hundred per cent. That means, if she took a part-time job to earn some extra cash, she'd immediately lose it in reduced benefits. Many other countries have marginal tax rates for low-income people that are well over 50 per cent, strongly discouraging them from working. Such a 'welfare trap' hardly seems sensible.

But it's also plausible to think that welfare states can *improve* economic productivity. If you lose your job, unemployment benefit means you don't have to rush into another one: it gives you time to find a new position that makes best use of your skills. Entrepreneurs might take more risks when they know that a bankruptcy won't be catastrophic: they could still send their kids to school, and get treatment when they're sick. In general, healthy, educated workers tend to be more productive. Sometimes, handouts can help in unexpected ways: in South Africa, girls grew up healthier when their grandmothers started getting pensions.

So: do welfare states boost economic growth or stunt it? That's not an easy question to answer – the systems have many moving parts, and each part could affect growth in many ways. But the weight of evidence suggests that it's a wash – the positive and negative effects balance out. Welfare states don't make the pie bigger or smaller, but they do change the size of each individual's slice. And that helps to keep a lid on inequality.

At least, it used to. In the last two decades, the data show welfare states haven't been doing that so well. Inequality, which in many countries widened sharply during the 1980s and 1990s, may widen further. And welfare states are creaking under the weight of a rapidly changing world.

There's demographic change: people are living for longer after retirement. There's social change: entitlements often date from an age when most women relied on male bread-winners, and most jobs were full-time and long-lasting. In the UK, for example, more than half the new jobs created since the 2007–8 financial crisis are in the self-employed sector. Yet a builder who's employed will get 'statutory sick pay' if he or she has an accident at work; a self-employed builder will not.

And there's globalisation: welfare states originated when employers were more geographically rooted than today's footloose multinationals – they couldn't easily relocate to jurisdictions with less burdensome regulations and taxes. Mobility of labour creates headaches, too – outraged news stories about immigrants claiming benefits arguably helped set Britain on the path to Brexit.

As we ponder how – or even whether – to fix the welfare state, we shouldn't forget that one of the biggest ways in which it shaped the modern economy was to take the heat out of demands for much more radical change.

Otto von Bismarck was no social reformer in the Frances Perkins mould. His motives were defensive. Bismarck feared that the public would turn to the revolutionary ideas of Karl Marx and Friedrich Engels; he hoped his welfare provisions would be just generous enough to keep the public quiescent. And that's a time-honoured political tactic – when the Roman emperor Trajan distributed free grain, the poet Juvenal famously grumbled that citizens could be bought off by 'bread and circuses'. You could tell much the same story about Italy's welfare state, which took shape in the 1930s as the fascist Mussolini tried to undercut the popular appeal of his socialist opponents.

In America, the New Deal was attacked as much from the left as the right. The populist Louisiana governor Huey Long complained that Frances Perkins hadn't gone far enough: he prepared to run for president on the slogan 'Share Our Wealth', and a promise to confiscate fortunes from the rich. He was shot dead, so that policy was never tested. At the beginning of the twenty-first century, such political tumult would have felt very distant. But now raw populist politics is back in many parts of the western world.

Perhaps we shouldn't be surprised by that. As we've seen, technological change has always created winners and losers, and the losers can always turn to politics if they're unhappy with how things are working out. In many industries, digital technologies are acting like modern-day gramophones, widening the gap between the top 0.1 per cent and the rest. Thanks to the power of search and seller feedback systems, new platforms are giving freelancers access to new markets. Or are they really freelancers? One of the pressing debates of the age is the extent to which Uber drivers or TaskRabbit taskers should be treated as employees – a status which in many countries unlocks access to parts of the welfare state.

The welfare state sits uneasily with large-scale international migration. People who instinctively feel that society should take care of its poorest members often feel very differently if those poorest members are migrants. But the interface between these two massive government endeavours – the welfare state and passport control – is often clunky. We should design our welfare states to fit snugly with our border controls, but we usually don't.

And the biggest question mark of all is whether, at long last, robots and artificial intelligence really will make large numbers of people completely unemployable. If human labour is less needed in future, that in principle is excellent news: a paradise of robotic servants awaits us. But our economies have always relied on the idea that people provide for themselves by selling their labour. If the robots make that impossible, then societies will simply come apart unless we reinvent the welfare state.

Not all economists think that's worth worrying about just yet. But those who do are reviving an idea that dates back to Thomas More's 1516 book *Utopia*: a universal basic income. The idea does seem utopian, in the sense of fantastically unrealistic: could we really imagine a world in which every-one gets a regular cash handout, enough to meet their basic needs, no questions asked?

Some evidence suggests it's worth considering. In the 1970s, the idea was trialled in a Canadian town called Dauphin – for years, thousands of residents got cheques every month. And it turns out that guaranteeing people an income had interesting effects. Fewer teenagers dropped out of school. Fewer people were hospitalised with mental health problems. And hardly anyone gave up work. New trials are underway, to see if the same thing happens elsewhere.

It would, of course, be enormously expensive. Suppose you gave every American adult, say, twelve thousand dollars a year. That would cost 70 per cent of the entire federal budget. It seems impossibly radical. But then, impossibly radical things do sometimes happen, and quickly. In the 1920s, not a single US state offered old-age pensions; by 1935, Frances Perkins had rolled them out across the nation.

II

REINVENTING HOW WE LIVE

My family's weekend newspaper wasn't complete without the *Innovations* catalogue – a glossy mail-order brochure advertising magnificently pointless products such as the 'Breath Alert' bad-breath detector or a necktie that operated via a concealed zip-fastener. *Innovations* eventually became defunct, to be replaced by Facebook ads for much the same junk. But while the brochure was absurd, the idea behind it has long been a tempting one: innovations are things that you can buy – or a better, cheaper way to make things that you've been buying all along.

It's not hard to see why this view is more attractive than a zip-up tie. It puts innovation into a box – perhaps even a gift-wrapped box. If innovation is about cool new things to enjoy, then it hardly seems threatening. If you don't want the cool new thing, you don't have to buy it, although there are plenty of advertisers who'd love to try to sell it to you.

But as we've already seen, inventions in the wild aren't quite so tame and cuddly. The plough was a better way to grow crops, but it wasn't just that: it ushered in an utterly new way of life, whether you personally used a plough or not. More recent innovations have the same quality. Collectively they have changed how we eat, how we play, how we care for children, where we live, and with whom we have sex. And these social changes have been intimately bound up with economic changes too – in particular, who gets paid a serious salary for their work and who gets paid nothing at all.

Real innovations don't come in a glossy brochure: they shape our world whether we buy them or not.

9

Infant Formula

It sounded like cannon fire. But where was it coming from? Pirates, probably. The *Benares*, a ship of the British East India Company, was docked at Makassar, on the Indonesian island of Sulawesi. Its commander gave the order to set sail and hunt them down.

Hundreds of kilometres away, on another Indonesian island – Java – soldiers in Yogyakarta heard the cannon noises, too. Their commander assumed the nearest town was under attack, so he marched his men there at once. But they found no cannons – just other people wondering what the noise was. Three days later, the *Benares* still hadn't found any pirates.

What they'd heard was the eruption of a volcano called Mount Tambora. When you consider that Mount Tambora is a thousand-odd kilometres from Yogyakarta, it's hard to imagine how terrifying the explosions must have been up close. A cocktail of toxic gas and liquefied rock roared down the volcano's slopes at the speed of a hurricane. It killed thousands. Mount Tambora was left four thousand feet shorter.

The year was 1815. Slowly, a vast cloud of volcanic ash drifted across the northern hemisphere, blocking the sun.

In Europe, 1816 became 'the year without a summer'. Crops failed, and desperate people ate rats, cats and grass. In the German town of Darmstadt, the suffering made a deep impression on a thirteen-year-old boy. Young Justus von Liebig loved helping out in his father's workshop, concocting pigments, paints and polishes for sale. He grew up to be a chemist, among the most brilliant of his age. And he was driven by the thought of making discoveries that might help prevent hunger. Liebig did some of the earliest research into fertilisers. He pioneered nutritional science – the analysis of food in terms of fats, proteins and carbohydrates. He invented beef extract.

Liebig invented something else, too: infant formula. Launched in 1865, Liebig's Soluble Food for Babies was a powder comprising cow's milk, wheat flour, malt flour and potassium bicarbonate. It was the first commercial substitute for breastmilk to come from rigorous scientific study.

As Liebig knew, not every baby has a mother who can breastfeed. Indeed, not every baby has a mother: before modern medicine, about one in a hundred childbirths killed the mother; it's little better in the poorest countries today. Then there are mothers who just can't make enough milk – the figures are disputed, but could be as high as one in twenty.

What happened to those kids before formula? Parents who could afford it employed wet-nurses – a respectable profession for the working girl, and an early casualty of Liebig's invention. Some used a goat or donkey. Many gave their infants 'pap', a bread-and-water mush, from hard-to-clean receptacles that must have teemed with bacteria. No wonder death rates were high: in the early 1800s, only two in three babies who weren't breastfed lived to see their first birthday.

Liebig's formula hit the market at a propitious time. Germ theory was increasingly well understood; the rubber teat had just been invented, too. The appeal of formula quickly spread beyond women who couldn't breastfeed. *Liebig's Soluble Food for Babies* democratised a lifestyle choice that had previously been open only to the well-to-do.

It's a choice that now shapes the modern workplace. For many new mothers who want – or need – to get back to work, formula is a godsend. And women are right to worry that taking time out might damage their careers. Recently, economists studied the experiences of the high-powered men and women emerging from Chicago University's MBA programme, entering the worlds of consulting and high finance. At first the women had similar experiences to the men, but over time a huge gap in earnings opened up. The critical moment? It was motherhood. Women took time off, and employers paid them less in response. Ironically, the men in the study were more likely than the women to have children. They just didn't change their working patterns.

There are both biological and cultural reasons why women are more likely than men to take time off when they start families. We can't change the fact that only women have wombs,* but we can try to change workplace culture. More governments are following Scandinavia's lead by giving dads the legal right to take time off. More leaders, like Facebook's Mark Zuckerberg, are setting the example of encouraging them to take it. And formula milk makes it a whole lot easier for dad to take over while mum gets back to work. Sure, there's also the option of the breast pump. But many mothers seem to find it a lot more trouble than formula: studies show that the less time mums have off work, the less likely they are to persevere with breastfeeding. That's hardly surprising.

---

* Strictly speaking, medical science now makes this possible. But it is unlikely to become a widely embraced option.

There's just one problem. Formula's not all that good for kids.

That shouldn't be surprising, either. Evolution, after all, has had thousands of generations to optimise the recipe for breastmilk. And formula doesn't quite match it. Formula-fed infants get sick more often. That leads to costs for medical treatment, and to parents taking time off work. It also leads to deaths, particularly in poorer countries where safe water sources aren't easy to come by. One credible estimate is that increased breastfeeding rates could save 800,000 children's lives each year. Justus von Liebig wanted his formula to save lives; he'd have been horrified.

Formula has another, less obvious economic cost: there's evidence that breastfed babies grow up with slightly higher IQs – around three points, when you control as best you can for other factors. What might be the benefit of making a whole generation of kids just that little bit smarter? According to *The Lancet*, about 300 billion dollars a year. That's several times the value of the global formula market.

Consequently, many governments try to promote breast-feeding. But nobody makes a quick profit from that. Selling formula, on the other hand, can be lucrative. Which have you seen more of recently: public service announcements about breastfeeding, or formula ads?

Those ads have always been controversial, not least because formula is arguably more addictive than tobacco or alco-hol: when a mother stops breastfeeding, her milk dries up. There's no going back. Liebig himself never claimed that his Soluble Food for Babies was better than breastmilk: he simply said he'd made it as nutritionally similar as possible. But he quickly inspired imitators who weren't so scrupulous. By the 1890s, adverts for formula routinely portrayed it as state of the art; meanwhile, paediatricians were starting to notice higher rates of scurvy and rickets among the offspring of mothers whom the advertising swayed.

The controversy peaked in 1974, when the campaigning group War on Want published a pamphlet called 'The Baby Killer' about how Nestlé sold infant formula in Africa; the boycotts lasted for years. By 1981, there was an 'International Code of Marketing Breast-Milk Substitutes'. But it's not hard law, and many say it's widely flouted. And there was fresh scandal in China in 2008, when industrial chemicals were found in formula milk and 300,000 children fell ill and some died.

What if there was a way to get the best of all worlds: equal career breaks for mums and dads, and breastmilk for infants, without the inconvenience of breast pumps? Perhaps there is – if you don't mind taking market forces to their logical conclusion. In Utah, there's a company called Ambrosia Labs. It pays mums in Cambodia to express breastmilk, screens it for quality, and sells it on to American mums. It's pricey now – over a hundred dollars a litre. But that could come down with scale. Governments might even be tempted to tax formula milk to fund a breastmilk market subsidy. Justus von Liebig sounded the death knell for wet nursing as a profession; perhaps the global supply chain is bringing it back.

10

TV Dinners

It's a typical November Tuesday for Mary, who lives in the northeast of the United States. She's forty-four, has a degree, and her family is prosperous – in the top quarter of American households by income. So, what has she been doing with her day? Is she a lawyer? A teacher? A management consultant?

No. Mary spent an hour knitting and sewing, two hours setting the table and doing the dishes, and well over two hours preparing and cooking food. In this, she isn't unusual. This is because it's 1965, and in 1965, many married American women – even those with an excellent education – spent large chunks of their day catering for their families. For these women, 'putting food on the table' wasn't a metaphor. It was something that they did quite literally – and it took many hours each week.

We know about Mary's day – and the days of many other people – because of time-use surveys conducted around the world. These are diaries of exactly how different sorts of people use their time. And for educated women, the way time is spent in the United States and other rich countries has changed radically over the past half a century. Women in the US now spend around forty-five minutes per day in total on cooking and cleaning up; that is still much more than men, who spend just fifteen minutes a day. But it is a vast change from Mary's four hours a day.

The reason for this shift is because of a radical change in the way the food we eat is prepared. If you want a symbol of this change, it's the introduction, in 1954, of the TV dinner. Presented in a space-age aluminium tray, and made so that the meat and the vegetables would all require the same cooking time, the 'frozen turkey tray TV dinner' was developed by a bacteriologist called Betty Cronin. She worked for the Swanson food processing company, which was looking for ways to keep busy after the business of supplying rations to US troops had dried up. Cronin herself, as an ambitious young career woman, was part of the ideal target market – women who were expected to cook for their husbands, yet were busy trying to develop their own careers. But she resisted the temptation: 'I've never had a TV dinner in my home,' she said in a 1989 interview. 'I used to work on them all day long. That was enough.'

But women didn't have to embrace the full aluminium-foil TV dinner experience to be liberated by changes in food processing. They had the freezer, the microwave, preservatives and production lines. Food had been perhaps the last cottage industry; it was something that would overwhelmingly be produced in the home. But food preparation has increasingly been industrialised – it's been outsourced to restaurants and takeaways, to sandwich shops, and to factories that prepare ready-to-eat or ready-to-cook meals. And the invention of the industrial meal – in all its forms – has led to a profound shift in the modern economy.

The most obvious symptom is that spending on food is

changing. American families spend more and more outside the home – on fast food, restaurant meals, sandwiches and snacks. Only a quarter of food spending was outside the home in the 1960s. It's been rising steadily over time, and in 2015 a landmark was reached: for the first time in their history, Americans spent more on food and drink outside the home than at grocery stores. In case you think the Americans are unusual in that, the British passed that particular milestone over a decade before.

Even within the home, food is increasingly processed to save the chef time, effort and skill. There are obvious examples where an entire meal comes ready to cook – a frozen pizza or one of Betty Cronin's one-tray creations. But there are less obvious ones, too – chopped salad in bags, meatballs or kebab sticks doused in sauce and ready to grill, pre-grated cheese, jars of pasta sauce, tea that comes packaged in an individual permeable bag, and chicken that comes plucked, gutted and full of sage and onion stuffing mix. There is even, of course, sliced bread. Each new innovation would seem bizarre to the older generation, but I've never plucked a chicken myself, and perhaps my children will never chop their own salads. All this saves time – serious amounts of time.

Such innovations didn't begin with the TV dinner: they've been a long time in the making. Households were buying pre-milled flour in the early nineteenth century, rather than having to take their own grains to a mill, or pound them into flour at home. In 1810 the French inventor Nicholas Appert patented a process for sealing and heat-treating food to preserve it. Condensed milk was patented in 1856; H.J. Heinz started to sell pre-cooked macaroni in the 1880s.

But these innovations didn't, at first, have an impact on how much time women spent preparing food. When the economist Valerie Ramey compared time use diaries in the

United States between the 1920s and the 1960s, she found an astonishing stability. Whether women were uneducated and married to farmers, or highly educated and married to urban professionals, they still spent a similar amount of time on housework, and that time did not change much for fifty years. It was only in the 1960s that the industrialisation of food really started to have a noticeable impact on the amount of housework that women did.

But surely the innovation responsible for emancipating women wasn't the frozen pizza, but the washing machine? The idea is widely believed, and it's appealing. A frozen TV dinner doesn't really feel like progress, compared to healthy home-cooked food. But a washing machine is clean and efficient and replaces work that was always drudgery. A washing machine is a robot washerwoman in cuboid form. It works. How could it not have been revolutionary?

It was, of course. But the revolution wasn't in the lives of women. It was in how lemon fresh we all started to smell. The data are clear that the washing machine didn't save a lot of time, because before the washing machine we didn't wash clothes very often. When it took all day to wash and dry a few shirts, people would use replaceable collars and cuffs or dark outer layers to hide the grime. But we cannot skip many meals in the way that we can skip the laundry. When it took two or three hours to prepare a meal, that was a job that someone had to take the time to do. The washing machine didn't save much time, and the ready meal did, because we were willing to stink but we weren't willing to starve.

The availability of ready meals has had some regrettable side effects. Obesity rates rose sharply in developed countries between the 1970s and the early twenty-first century, at much the same time as these culinary innovations were being developed. This is no coincidence, say health economists: the cost of eating a lot of calories has fallen dramatically, not just in financial terms but in terms of time.

Consider the humble potato. It has long been a staple of the American diet, but before the Second World War potatoes were usually baked, mashed or boiled. There's a reason for that: roast potatoes need to be peeled, chopped, parboiled and then roasted; French fries or chips must be finely chopped and then deep fried. This is all time consuming.

Over time, however, the production of fried sliced potato chips – both French fries and crisps – was centralised. French fries can be peeled, chopped, fried and frozen in a factory; they are then refried in a fast-food restaurant or microwaved at home. Between 1977 and 1995, American potato consumption increased by a third, and that increase was almost entirely explained by the rise of fried potatoes.

Even simpler, crisps can be fried, salted, flavoured and packaged to last for many weeks on the shelf. But this convenience comes at a cost. In the USA, calorie intake by adults rose by about 10 per cent between the 1970s and the 1990s; but none of that was as a result of more calorific regular meals. It was all snacking – and that usually means processed convenience food.

Psychology – and common sense – suggest this shouldn't be a surprise. Experiments conducted by behavioural scientists show that we make very different decisions about what to eat depending on how far away the meal is. A long-planned meal is likely to be nutritious, but when we make more impulsive decisions our snacks are more likely to be junk food than something nourishing.

The industrialisation of food – symbolised by the TV dinner – changed our economy in two important ways. It freed women from hours of domestic chores, removing a large obstacle to them adopting serious professional careers.

But by making empty calories ever more convenient to acquire, it also freed our waistlines to expand. The challenge now – as with so many inventions – is to enjoy the benefit without also suffering the cost.

11

The Pill

Infant formula changed what it meant to be a mother, and TV dinners changed what it meant to be a housewife. But the contraceptive pill changed both – and more besides. It had profound social consequences, and indeed that was the point – at least in the view of Margaret Sanger, the birth-control activist who urged scientists to develop it. Sanger wanted to liberate women sexually and socially, to put them on a more equal footing with men.

But the pill wasn't just socially revolutionary. It also sparked an economic revolution – perhaps the most significant economic change of the late twentieth century.

To see why, first consider what the pill offered to women.

For a start, it worked – which is more than you can say for many of the alternatives. Over the centuries, lovers have tried all kinds of unappealing tricks to prevent pregnancy. There was crocodile dung in ancient Egypt, Aristotle's recommendation of cedar oil, and Casanova's method of using half a lemon as a cervical cap. But even the obvious modern alternative to the pill, condoms, have a failure rate. Because people don't tend to use condoms exactly as they're supposed to, they sometimes rip or slip – with the result that for every hundred sexually active women using condoms for a year, eighteen will become pregnant. Not great. The failure rate of the sponge is similar; the diaphragm isn't much better.

But the failure rate of the pill is just 6 per cent – three times safer than condoms. In fact, that assumes typical use – use it perfectly and the failure rate drops to one twentieth of that. And responsibility for using the pill perfectly was the woman's, not that of her fumbling partner.

The pill gave women control in other ways. Using a condom meant negotiating with a partner. The diaphragm and sponge were messy. But the decision to use the pill was a woman's, and it was private. The pill was neat and it was discreet. No wonder women wanted it. The pill was first approved in the United States in 1960, and it took off almost immediately – in just five years, almost half of married women on birth control were using it.

But the real revolution would come when *unmarried* women could use oral contraceptives. That took time. But around 1970, ten years after the pill had been approved, state after state in the US was making it easier for young unmarried women to get the pill. Universities began to open family planning centres, and their female students began to use them. By the mid-1970s, the pill was overwhelmingly the most popular form of contraception for eighteen- and nineteen-year-old women in the US.

That was when the economic revolution also began. Beginning in America in the 1970s, women started studying particular kinds of degrees: law, medicine, dentistry and MBAs. These degrees had been very masculine until then. In 1970, men earned over 90 per cent of the medical degrees awarded that year. Law degrees and MBAs were over 95 per cent male. Dentistry degrees were 99 per cent male. But at the beginning of the 1970s, equipped with the pill, women surged into all these courses. The proportion of women in these classes increased swiftly, and by 1980 they were often a third of the class. It was a huge change in a brief space of time.

This wasn't simply because women were more likely to go to university. Women who'd already decided to be students were opting for these professional courses. The proportion of all female students studying subjects such as medicine and law rose dramatically, and logically enough, the presence of women in the professions rose sharply shortly afterwards.

But what did this have to do with the pill?

The answer is that by giving women control over their fertility, the pill allowed them to invest in their careers. Before the pill was available, taking five years or more to qualify as a doctor or a lawyer did not look like a good use of time and money if pregnancy was a constant risk. To reap the benefits of those courses, a woman would need to be able to reliably delay becoming a mother until she was thirty at least – having a baby might derail her studies or delay her professional progress at a critical time. A sexually active woman who tried to become a doctor, dentist or lawyer was doing the equivalent of building a factory in an earthquake zone: just one bit of bad luck and the expensive investment might be trashed.

Of course, women could have simply abstained from sex if they wanted to study for a professional career. But many of them didn't want to. And that decision wasn't just about having fun; it was also about finding a husband. Before the pill, people married young. A woman who decided to abstain from sex while developing her career might try to find a husband at the age of thirty and find that, quite literally, all the good men had been taken.

The pill changed both those dynamics. It meant that unmarried women could have sex with substantially less risk of an unwanted pregnancy. But it also changed the whole pattern of marriage. Everyone started to marry later. Why hurry? And that meant that even women who didn't use the pill found that they didn't have to rush into marriage either. The babies started to arrive later, and at a time that women chose for themselves. And that meant that women, at least, had time to establish a professional career.

Of course, many other things were changing for American women in the 1970s. There was the legalisation of abortion around the same time, laws against sex discrimination, the social movement of feminism and the fact that young men were being drafted to fight in Vietnam, leaving employers keen to recruit women in their place.

But a careful statistical study by the Harvard economists Claudia Goldin and Lawrence Katz strongly suggests that the pill must have played a major role in allowing women to delay marriage, delay motherhood and invest in their own careers. When you look at the other factors that were changing, the timing isn't quite right to explain what happened. But when Goldin and Katz tracked the availability of the pill to young women, state by state, they found that as each state opened up access to the technology, the enrolment rate in professional courses soared – and so did women's wages.

A few years ago, an economist called Amalia Miller used a variety of clever statistical methods to demonstrate that if a woman in her twenties was able to delay motherhood by one year, her lifetime earnings would rise by 10 per cent: that was some measure of the vast advantage to a woman of completing her studies and securing her career before having children. But the young women of the 1970s didn't need to see Amalia Miller's research: they already knew it was true. As the pill became available, they signed up for long professional courses in undreamt-of numbers.

American women today can look across the Pacific Ocean for a vision of an alternative reality. In Japan, one of the world's most technologically advanced societies, the pill wasn't approved for use until 1999. Japanese women had to wait thirty-nine years longer than American women for the same contraceptive; in contrast, when the erection-boosting drug Viagra was approved in the US, Japan was just a few months behind. Gender inequality in Japan is generally reckoned to be worse than anywhere else in the developed world, with women continuing to struggle for recognition in the workplace. It is impossible to disentangle cause and effect here, but the experience in the US suggests that it is no coincidence – delay the pill by two generations, and the economic impact on women will be enormous. It is a tiny little pill that continues to transform the world economy.

12

Video Games

E arly in 1962, a young student at the Massachusetts Institute of Technology (MIT) was on his way to his home in the nearby town of Lowell. It was a cold night, with a cloudless sky, and as Peter Samson stepped off the train and gazed up at the starfield, a meteor streaked across the heavens. But instead of gasping at the beauty of creation, Samson reflexively grabbed for a game controller that wasn't there, and scanned the skies, wondering where his spaceship had gone. Samson's brain had grown out of the habit of looking at the real stars. He was spending way too much time playing Spacewar.

Samson's near hallucination was the precursor of countless digital fever dreams to come – that experience of drifting off to sleep dreaming of Pac Man, or rotating Tetris blocks, or bagging a rare Pokemon Jigglypuff. Or, for that matter, the reflexive checking of your phone for the latest Facebook update. That ability of a computer to yank our Pavlovian reflexes and haunt our sleep – in 1962, that would have been unimaginable to anyone but Peter Samson and a few of his hacker friends. They were avid players of Spacewar, the first video game that mattered – the one that opened the door to a social craze and a massive industry, and shaped our economy in more profound ways than we realise.

Before Spacewar, computers were intimidating: large grey cabinets in purpose-built rooms, closed off to all but the highly trained. Vast and expensive, forbidding and corporate. Computing was what banks did, and corporations, and the military: computers worked for the suits.

But at the beginning of the 1960s, at MIT, new computers were being installed in a more relaxed environment. They didn't have their own rooms – they were part of the laboratory furniture. Students were allowed to mess around with them. The term 'hacker' was born, and rather than having the modern mass-media sense of a malevolent cracker of security systems, it meant someone who would experiment, cut corners, produce strange effects. And just as hacker culture was being born, MIT ordered a new kind of computer: the PDP-1. It was compact – the size of a large fridge – and relatively easy to use. It was powerful. And – oh joy! – it communicated not through a printer, but through a high-precision cathode ray tube. A video display.

When a young researcher called Slug Russell heard about the PDP-1, he and his friends began plotting the best way to show off its capabilities. They had been reading a lot of science fiction. They had been dreaming of a proper Hollywood space opera – this was nearly two decades before *Star Wars*. But since no such movie was in the offing, they plumped for the best possible alternative: Spacewar – a two-player video game that pitted starship captains against each other in a photon-torpedo-powered duel to the death.

There were two ships – just a few pixels outlining the starcraft – and the players could spin, thrust, or fire torpedoes. Other enthusiasts soon joined in, making the game smoother and faster, adding a star with a realistic gravitational pull, and cobbling together special controllers from plywood, electrical toggles and Bakelite. They were hackers, after all.

One of them decided that Spacewar deserved a breathtaking backdrop, and programmed what he called the 'Expensive Planetarium' sub-routine. It displayed a realistic starscape – stars shown with five different brightnesses – as viewed from the Earth's equator. The author of the glorious addition: Peter Samson, the young student whose imagination was so captured by Spacewar that he misperceived the night sky above Lowell.

In one way, the economic legacy of Spacewar is obvious. As computers became cheap enough to install in arcades, and then in the home, the games industry blossomed. One of the early hits, Asteroids, owed a clear debt to Spacewar – with the realistic-seeming physics of a spaceship that rotated and thrust in a zero-gravity environment. Computer games now rival the film industry for revenue. They're becoming culturally important, too: Lego's Minecraft tie-in jostles for popularity with their Star Wars and Marvel sets.

But beyond the money that we spend on them, games affect the economy in a couple of ways. First, virtual worlds can create real jobs. One of the first people to make this case was an economist named Edward Castronova. In 2001 Castronova calculated the gross national product per capita of an online world called Norrath – the setting for an online role-playing game, Everquest. Norrath wasn't particularly populous – around 60,000 people would be logged in at a time, performing mundane tasks to accumulate treasure which they could use to buy enjoyable capabilities for their characters. Except that some players were impatient. They bought virtual treasure from other players, on sites like eBay, for real money. Which meant other players could earn real money for doing mundane work in Norrath.

The wage, reckoned Castronova, was about $3.50 an hour – not much for a Californian but an excellent rate if you happened to live in Nairobi. Before long, 'virtual sweatshops' sprang up from China to India, where teenagers ground away on the tedious parts of certain games, acquiring digital short-cuts to sell to more prosperous players who wanted to get straight to the good stuff. And it still happens: some people are making tens of thousands of dollars a month on auction sites in Japan just selling virtual game characters.

For most people, though, virtual worlds aren't a place to earn money, but to enjoy spending time: cooperating in guilds, mastering complex skills, or having a party inside their own imaginations. Even as Castronova was writing about tiny Norrath, 1.5 million South Koreans were playing in the virtual world of a game called Lineage. Then came Farmville on Facebook, blurring a game with a social network; and mobile games, like Angry Birds and Candy Crush Saga; and augmented reality games, like Pokemon Go. By 2011, the game scholar Jane McGonigal estimated that more than half a billion people worldwide were spending serious amounts of time – almost two hours a day, on average – playing computer games. A billion or two is within easy reach.

And that brings us to the second economic impact. How many of those people are choosing virtual fun over boring work for real money?

A decade ago, I saw Edward Castronova speak in front of a learned audience of scientists and policy wonks in Washington DC. You guys are already winning in the game of real life, he told us. But not everyone can. And if your choice is to be a Starbucks server or a starship captain – what, really, is so crazy about deciding to take command in an imaginary world?

Castronova may have been onto something. In 2016, four

economists presented research into a puzzling fact about the US labour market: the economy was growing strongly and unemployment rates were low, and yet a surprisingly large number of able-bodied young men were either working part-time or not working at all. More puzzling still, while most studies of unemployment find that it makes people thoroughly miserable, against expectations the happiness of these young men was rising. The researchers concluded that the explanation was – well, living at home, sponging off their parents, and playing video games. These young men were deciding they didn't want to be a Starbucks server. Being a starship captain was far more appealing.

13

Market Research

In the early years of the twentieth century, US car makers had it good. As quickly as they could manufacture cars, people bought them. By 1914, that was changing. In higher price brackets, especially, purchasers and dealerships were becoming choosier. One commentator warned that the retailer 'could no longer sell what his own judgement dictated. He must sell what the consumer wanted.'

That commentator was Charles Coolidge Parlin. He's widely recognised as the world's first professional market researcher – and, indeed, as the man who invented the very idea of market research. A century later, the market research profession is huge: in the United States alone, it employs around half a million people.

Parlin was tasked with taking the pulse of the US automobile market. He travelled tens of thousands of miles, and interviewed hundreds of car dealers. After months of work, he presented his employer with what he modestly described as '2500 typewritten sheets, charts, maps, statistics, tables etc'.

You might wonder which car maker employed Parlin to conduct this research. Was it, perhaps, Henry Ford, who at the time was busy gaining an edge on his rivals with another innovation – the assembly line?

But no: Ford didn't have a market research department to gauge what customers wanted. Perhaps that's no surprise. Henry Ford is widely supposed to have quipped that people could have a Model T in 'any colour they like, as long as it's black'.

In fact, no car makers employed market researchers. Parlin had been hired by a magazine publisher. The Curtis Publishing Company published some of the most widely read periodicals of the time: the *Saturday Evening Post*, *The Ladies' Home Journal* and *The Country Gentleman*. The magazines depended on advertising revenue. The company's founder thought he'd be able to sell more advertising space if advertising were perceived as more effective, and he wondered if researching markets might make it possible to devise more effective adverts. In 1911, he set up a new division of his company to explore this vaguely conceived idea.

The first head of that research division was Charles Coolidge Parlin. It wasn't an obvious career move for a 39-year-old high school principal from Wisconsin – but then, being the world's first market researcher wouldn't have been an obvious career move for anyone. Parlin started by immersing himself in agricultural machinery; then he tackled department stores. Not everyone saw value in his activities, at first – even as he introduced his pamphlet 'The merchandising of automobiles; an address to retailers', he still felt the need to include a diffident justification of his job's existence. He hoped to be 'of constructive service to the industry as a whole', he wrote, explaining that car makers spent heavily on advertising and his employers wanted to 'ascertain whether this important source of business was one which would continue'.

The invention of market research marks an early step in

a broader shift from a 'producer-led' to a 'consumer-led' approach to business – from making something then trying to persuade people to buy it, to trying to find out what people might buy and then making it.

The producer-led mindset is exemplified by Henry Ford's 'any colour, as long as it's black'. From 1914 to 1926, only black Model Ts rolled off Ford's production line: it was simpler to assemble cars of a single colour, and black paint was cheap and durable. All that remained was to persuade customers that what they really wanted was a black Model T. To be fair, Ford excelled at this.

Few companies nowadays would simply produce what's convenient then hope to sell it. A panoply of market research techniques – surveys, focus groups, beta testing – helps determine what might sell. If metallic paint and go-faster stripes will sell more cars, that's what will get made.

Where Parlin led, others eventually followed. By the late 1910s, not long after his report on automobiles, companies had started setting up their own market research departments. Over the next decade, US advertising budgets almost doubled. Approaches to market research became more scientific: in the 1930s, George Gallup pioneered opinion polls; the first focus group was conducted in 1941 by an academic sociologist, Robert K. Merton. He later wished he could have patented the idea and collected royalties.

But systematically *investigating* consumer preferences was only part of the story; marketers also realised it was possible systematically to *change* them. Merton coined a phrase to describe the kind of successful, cool or savvy individual who routinely features in marketing campaigns. That phrase: the 'role model'.

The nature of advertising was changing: no longer merely providing information, but trying to manufacture desire.

Sigmund Freud's nephew, Edward Bernays, pioneered the fields of public relations and propaganda. Among his most famous stunts for corporate clients, in 1929 Bernays helped the American Tobacco Company to persuade women that smoking in public was an act of female liberation. Cigarettes, he said, were 'torches of freedom'.

Today, attempts to discern and direct public preferences shape every corner of the economy. Any viral marketer will tell you that creating buzz remains more of an art than a science; but with ever more data available, investigations of consumer psychology can get ever more detailed. Where Ford offered cars in a single shade of black, Google famously tested the effect on click-through rates of forty-one slightly different shades of blue.

Should we worry about the reach and sophistication of corporate efforts to probe and manipulate our consumer psyches? An evolutionary psychologist called Geoffrey Miller takes a more optimistic view: 'Like chivalrous lovers,' Miller writes, 'the best marketing-oriented companies help us discover desires we never knew we had, and ways of fulfilling them we never imagined.' Perhaps.

Miller sees humans showing off through our consumer purchases much as peacocks impress peahens with their tails; such ideas hark back to an economist and sociologist named Thorstein Veblen. Veblen invented the concept of conspicuous consumption back in 1899.

Parlin had read his Veblen. He understood the signalling power of consumer purchases: 'the pleasure car', he wrote in an address to retailers, 'is the traveling representative of a man's taste or refinement ... a dilapidated pleasure car, like a decrepit horse, advertises that the driver is lacking in funds or lacking in pride'. In other words, perhaps not someone you should trust as a business associate. Or as a husband.

Signalling these days is much more complex than merely displaying wealth: we might choose a Prius if we want to display our green credentials, or a Volvo if we want to be seen as safety-conscious. These signals carry meaning only because brands have spent decades consciously trying to understand and respond to consumer desires – and to shape them.

By contrast with today's adverts, those of 1914 were delightfully unsophisticated. The tagline of one, for a Model T: 'Buy it because it's a better car'. Isn't that advertisement, in its own adorable way, perfect? But it couldn't last. Charles Coolidge Parlin was in the process of ushering us towards a very different world.

14

Air Conditioning

If only we could control the weather – at the push of a button, making it warmer or cooler, wetter or drier. We'd have no more droughts or floods, no heatwaves or icy roads. Deserts would become verdant. Crops would never fail. And we could stop worrying about climate change. As it happens, the threat of climate change has sparked some crazy-sounding ideas for hacking the climate, like spraying sulphuric acid into the upper atmosphere to cool it down, or dumping quicklime in the oceans to absorb carbon dioxide and slow down the greenhouse effect. Other scientists, meanwhile, are working on realising the shaman's dream of making rain; their techniques include seeding clouds with silver iodide and sending electrically charged particles into the sky.

Clever as humans are, however, we're nowhere near precision control of the weather. At least if we're talking about *outside*. Since the invention of air conditioning, we can control the weather *inside*. That's not quite as big a deal, but it has still had some far-reaching and unexpected effects.

Ever since our ancestors mastered fire, humans have been able to get warmer when it's cold. Cooling down when it's hot has been more of a challenge. The eccentric teenage Roman emperor Elagabalus made an early attempt at air conditioning by sending slaves into the mountains to bring down snow and pile it in his garden, where breezes would carry the cooler air inside.

Needless to say, this was not a scalable solution. At least, not until the nineteenth century, when a Boston entrepreneur named Frederic Tudor amassed an unlikely fortune in a similar way. In 1806, he began carving blocks of ice from New England's frozen lakes in winter, insulating them in sawdust, and shipping them to warmer climes for summer. This was profitable for the rest of the century, and the warmer parts of the US became addicted to New England ice. Mild New England winters would cause panic about an 'ice famine'.

Air conditioning as we know it began in 1902, and it was nothing to do with human comfort. The workers at Sackett & Wilhelms Lithography and Printing Company in New York were frustrated with varying humidity levels when trying to print in colour. The process required the same paper to be printed four times – in cyan ink, magenta, yellow and black. If the humidity changed between print runs, the paper would slightly expand or contract. Even a millimetre's misalignment looked awful.

The printers asked Buffalo Forge, a heating company, if it could devise a system to control humidity. Buffalo Forge assigned the problem to a young engineer, just a year out of university. Willis Carrier was earning just ten dollars a week – below minimum wage in today's money. But he figured out a solution: circulating air over coils that were chilled by compressed ammonia maintained the humidity at a constant 55 per cent.

The printers were delighted, and Buffalo Forge was soon selling Willis Carrier's invention wherever humidity posed problems: from textiles to flour mills to the Gillette corporation, where excessive moisture rusted its razor blades. These early, industrial clients didn't much care about making temperatures more comfortable for their workers; that was an incidental benefit of controlling the humidity. But Carrier saw an opportunity. By 1906 he was already talking up the potential for 'comfort' applications in public buildings such as theatres.

It was an astute choice of target market. Historically, theatres often shut down for summer – on a stifling-hot day, nobody wanted to see a play. It's not hard to imagine why: no windows; human bodies tightly packed; and, before electricity, lighting provided by flares that gave off heat. New England ice was briefly popular: in the summer of 1880, New York's Madison Square Theatre used four tons of ice a day, and an eight-foot fan to blow air over the ice and through ducts towards the audience. Unfortunately, this wasn't an ideal solution. The air, though cool, was also damp, and pollution was increasing in New England's lakes. Sometimes, as the ice melted, into the auditorium wafted some unpleasant smells.

Willis Carrier called his system of cooling the 'weathermaker', and it was much more practical. The burgeoning movie theatres of the 1920s were where the general public first experienced air conditioning, and it quickly became as much of a selling point as the new talking films. The enduring and profitable Hollywood tradition of the summer blockbuster traces directly back to Carrier. So does the rise of the shopping mall.

But air conditioning has become more than a mere convenience. Computers fail if they get too hot or damp, so air conditioning enables the server farms that power the internet. Indeed, if factories couldn't control their air quality, we'd struggle to manufacture silicon chips at all.

Air conditioning is a revolutionary technology; it has had a profound influence on where and how we live. It has transformed architecture. Historically, a cool building in a hot climate implied thick walls, high ceilings, balconies, courtyards, and windows facing away from the sun. The dogtrot house, popular in America's south, was bisected by a covered, open-ended corridor to let breezes through. Glass-fronted skyscrapers were not a sensible option: you'd bake on the upper floors. With air conditioning, old workarounds become irrelevant and new designs become possible.

Air conditioning has changed demographics, too. Without it, it's hard to imagine the rise of cities like Houston, Phoenix or Atlanta, as well as Dubai or Singapore. As residential units spread rapidly across America in the second half of the twentieth century, population boomed in the 'sun belt' – the warmer south of the country, from Florida to California – from 28 per cent of Americans to 40 per cent. As retirees, in particular, moved from north to south, they also changed the region's political balance; the author Steven Johnson has plausibly argued that air conditioning elected Ronald Reagan.

Reagan became president in 1980: back then, America alone, with just 5 per cent of the world's population, used more than half the world's air conditioning. Emerging economies have since caught up quickly: China will soon become the global leader. The proportion of air-conditioned homes in Chinese cities jumped from under a tenth to more than two-thirds in just ten years. In countries like India, Brazil and Indonesia, the market for air conditioners is expanding at double-digit rates. And there's plenty more room for growth: from Manila to Kinshasa, eleven of the world's thirty largest cities are in the tropics.

The boom in air conditioning is good news for many reasons – beyond the obvious, that life in a hot, humid summer is simply more pleasant with it than without. Air conditioning lowers the death rate during heatwaves. In prisons, heat makes inmates fractious – air conditioning pays for itself by reducing fights. In exam halls, when the temperature exceeds the low twenties centigrade (low seventies Fahrenheit), students start to score lower in maths tests. In offices, air conditioning makes us more productive: according to one early study, it enabled US government typists to do 24 per cent more work.

Economists have since confirmed that there's a relationship between productivity and keeping cool. William Nordhaus of Yale divided the world into cells, by lines of latitude and longitude, and plotted each one's climate, output and population: the hotter the average temperature, he found, the less productive people could be. According to Geoffrey Heal of Columbia University and Jisung Park of Harvard, a hotter-than-average year is bad for productivity in hot countries, but good in cold ones: crunching the numbers, they conclude that human productivity peaks at between eighteen and twenty-two degrees.

But there's an inconvenient truth: you can only make a room or building cooler inside by making it warmer outside. Air conditioning units pump hot air out of buildings – a study in Phoenix, Arizona, found that this increased the city's night-time temperature by two degrees. Of course, that only makes air conditioning units work harder, making the outside hotter still. On underground metro systems, cooling the trains can lead to swelteringly hot platforms. Then there's the electricity that powers air conditioning, often made by burning gas or coal, and the coolants air conditioners use, many of which are powerful greenhouse gases when they leak.

You'd expect air conditioning technology to be getting cleaner and greener, and you'd be right. But demand is growing so quickly that, even if the optimists are right about possible efficiency gains, there'll be an eightfold increase in energy consumption by 2050. That's worrying news for climate change.

Will we get inventions to control the outdoor weather, too? Perhaps. But even air conditioning – a brilliant but simple and straightforward invention – has had some powerful and unexpected side effects. Controlling the climate itself will be neither simple nor straightforward. And the side effects? We can barely imagine.

15

Department Stores

'No, I'm just looking': these are words most of us have said, at some point, when browsing in a store and approached politely by a sales assistant. Most of us will not have then experienced the sales assistant snarling 'then 'op it, mate!'

Hearing those words in a London shop made quite an impression on Harry Gordon Selfridge. The year was 1888, and the flamboyant American was touring the great department stores of Europe – in Vienna and Berlin, the famous Bon Marché in Paris, and then Manchester and London – to see what tips he could pick up for his then-employer, Chicago's Marshall Field. Field had been busily popularising the aphorism 'the customer is always right'. Evidently, this was not yet the case in England.

Two decades later, Selfridge was back in London, opening his eponymous department store on Oxford Street – now a global mecca for retail, then an unfashionable backwater, but handily near a station on a newly opened underground line. Selfridge's caused a sensation. This was due partly to its sheer size – the retail space covered 6 acres. Plate glass windows had been a feature of high streets for a few decades, but Selfridge installed the largest glass sheets in the world – and he created, behind them, the most sumptuous shop window displays.

But more than scale, what set Selfridge's apart was attitude. Harry Gordon Selfridge was introducing Londoners to a whole new shopping experience, one honed in the department stores of late-nineteenth-century America.

'Just looking' was positively encouraged. As he had in Chicago, Selfridge swept away the previous shopkeepers' custom of stashing the merchandise in places where sales assistants had to fetch it for you – in cabinets, behind locked glass doors, or high up on shelves that could be reached only with a ladder. He instead laid out the open aisle displays we now take for granted, where you can touch a product, pick it up, and inspect it from all angles, without a salesperson hovering by your side. In the full-page newspaper adverts he took out when his store opened, Selfridge compared the 'pleasures of shopping' to those of 'sight-seeing'.

Shopping had long been bound up with social display: the old arcades of the great European cities displaying their fine cotton fashions – gorgeously lit with candles and mirrors – were places for the upper classes not only to see, but to be seen. Selfridge had no truck with snobbery or exclusivity. His adverts pointedly made clear that the 'whole British public' would be welcome – 'no cards of admission are required'. Management consultants nowadays talk about the fortune to be found at the 'bottom of the pyramid'; Selfridge was way ahead of them. In his Chicago store, he appealed to the working classes by dreaming up the concept of the 'bargain basement'.

Selfridge did perhaps more than anyone to invent shopping as we know it. But the ideas were in the air. Another trailblazer was an Irish immigrant named Alexander Turney

Stewart. It was Stewart who introduced New Yorkers to the shocking concept of not hassling customers the moment they walked through the door. He called this novel policy 'free entrance'.

A.T. Stewart and Co. was among the first stores to practise the now-ubiquitous 'clearance sale', periodically moving on the last bits of old stock at knockdown prices to make room for new. Stewart offered no-quibble refunds. He made customers pay in cash, or settle their bills quickly; traditionally, shoppers had expected to string out their lines of credit for up to a year.

Another insight Stewart applied at his store was that not everybody likes to haggle – some of us welcome the simplicity of being quoted a fair price, and told we can take it or leave it. Stewart made possible this 'one-price' approach by accepting unusually low mark-ups. '[I] put my goods on the market at the lowest price I can afford,' he explained, 'although I realize only a small profit on each sale, the enlarged area of business makes possible a large accumulation of capital.'

This idea wasn't totally unprecedented, but it was certainly considered radical. The first salesman Stewart hired was appalled to hear that he would not be allowed to apply his finely tuned skill of sizing up the customer's apparent wealth and extracting as extravagant a price as possible: he resigned on the spot, telling the youthful Irish shopkeeper he'd be bankrupt within a month. By the time Stewart died, over five decades later, he was one of the richest men in New York.

The great department stores became cathedrals of commerce: at Stewart's 'Marble Palace', the shopkeeper boasted 'You may gaze upon a million dollars' worth of goods, and no man will interrupt either your meditation or your admiration.' They took shopping to another level, sometimes literally: Corvin's in Budapest installed an elevator that became such an attraction in its own right that they began to charge for using it. Harrods in London had a moving staircase carrying four thousand people an hour.

In such shops one could buy anything from cradles to gravestones – Harrods offered a full funeral service including hearse, coffin and attendants. There were picture galleries, smoking rooms, tea rooms, concerts. The shop displays bled out into the street, as entrepreneurs built out covered galleries around their stores. It was, says Frank Trentmann, a historian, the birth of 'total shopping'.

The glory days of the city centre department store have faded a little. With the rise of cars has come the out-of-town shopping mall, where land is cheaper. Tourists in England still enjoy Harrods and Selfridges, but many also head to Bicester Village, a few miles north of Oxford, an outlet shopping centre that specialises in luxury brands at a discount.

But the experience of going to the shops has changed remarkably little since pioneers like Stewart and Selfridge turned it on its head. And it may be no coincidence that they did it at a time when women were gaining in social and economic power.

There are, of course, some tired stereotypes about women and their supposed love of shopping. But the data suggest that the stereotypes aren't completely imaginary. Time-use studies suggest women spend more time shopping than men do. Other research suggests that this is a matter of preference as well as duty: men tend to say they like shops with easy parking and short checkout lines, so they can get what they came for and leave; women are more likely to prioritise aspects of the shopping experience, like the friendliness of sales assistants.

Such research wouldn't have shocked Harry Gordon Selfridge. He saw that female customers offered profitable opportunities that other retailers were bungling, and made a point of trying to understand what they wanted. One of his quietly revolutionary moves: Selfridge's featured a ladies' lavatory. Strange as it may sound to modern ears, this was a facility London's shopkeepers had hitherto neglected to provide. Selfridge saw, as other men apparently had not, that women might want to stay in town all day, without having to use an insalubrious public convenience or retreat to a respectable hotel for tea whenever they wanted to relieve themselves.

Lindy Woodhead, who wrote a biography of Selfridge, even thinks he 'could justifiably claim to have helped emancipate women'. That's a big claim for any shopkeeper. But social progress can sometimes come from unexpected directions. And Harry Gordon Selfridge certainly saw himself as a social reformer. He once explained why, at his Chicago store, he'd introduced a crèche: 'I came along just at the time when women wanted to step out on their own,' he said. 'They came to the store and realized some of their dreams.'

III

INVENTING NEW SYSTEMS

Late in 1946, a group of engineers from more than two dozen nations gathered together in London. It wasn't the easiest of times or places to hold a conference. 'All the hotels were very good, but very short of supplies,' recalled Willy Kuert, a Swiss delegate. But he understood the difficulties. As long as you focused on the quality of the food rather than the quantity, there was no cause for complaint.

Kuert and his colleagues had a plan: they wanted to establish a new organisation that would agree international standards. Even amidst the wreckage of the war, the big tension was between those who measured with inches and those who measured with centimetres. 'We didn't talk about it,' said Kuert. 'We would have to live with it.' Despite that tension, it was a friendly atmosphere – people liked each other and wanted to get things done. And in due course, the conference reached an agreement: the establishment of the International Organization for Standardization, or ISO.

The ISO, of course, sets standards. Standards for nuts and bolts, for pipes, for ball-bearings, for shipping containers and for solar panels. Some of these standards are touchy-feely (standards for sustainable development) and some are cutting-edge (standards for hydrogen refuelling stations). But for old ISO hands, it's the humble stuff that counts: getting the UK to accept international standards on screw threads is still remembered as one of the ISO's great achievements. Alas, the ISO hasn't quite managed to standardise standard-setting bodies: it must instead rub along with the International Electrotechnical Commission and the International Telecommunication Union, and no doubt many more.

It's easy to chuckle at the idea of international standards for nuts and bolts – but then again, non-standardised nuts and bolts wouldn't be funny at all. From food labels that make sense to us to cars that start when we turn the key, from cell phones that can call other cell phones to power plugs that fit into power sockets, our modern economy is built on standardisation. There's no glory in standardised ball-bearings, but a smoothly functioning economy runs upon such things in both a metaphorical and a literal sense.

Many key inventions work only as part of a broader system. That system may be one of pure engineering standards – as with a cell phone. But it may also be a more human system. For example, paper money has no intrinsic value: it works only if people expect other people to accept the paper as payment. And an invention such as the elevator works much better when combined with other technologies: reinforced concrete to build skyscrapers; air conditioning to keep them cool; and public transport to deliver people to dense business districts.

But let's start with one of the most important inventions in human history – one that only began to realise its potential when all kinds of systems were adapted to fit around it.

16

The Dynamo

For investors in Boo.com, WebVan and eToys, the bursting of the dotcom bubble came as a bit of a shock. Companies like this raised vast sums on the promise that the World Wide Web would change everything. Then, in the spring of 2000, stock markets collapsed.

Some economists had long been sceptical about the promise of computers. In 1987, we didn't have the Web, but spreadsheets and databases were appearing in every workplace. And having, it seemed, no impact whatsoever. The leading thinker on economic growth, Robert Solow, quipped, 'you can see the computer age everywhere but in the productivity statistics'.

It's not easy to track the overall economic impact of innovation, but the best measure we have is something called 'total factor productivity'. When it's growing, that means the economy is somehow squeezing more output out of inputs, such as machinery, human labour and education. In the 1980s, when Solow was writing, it was growing at the slowest rate for decades – slower even than the Great Depression. Technology seemed to be booming, but productivity was almost stagnant.

Economists called it the 'productivity paradox'. What might explain it?

For a hint, rewind a hundred years. Another remarkable new technology was proving disappointing – electricity. Some corporations were investing in electric dynamos and motors, and installing them in the workplace. Yet the surge in productivity would not come.

The potential of electricity seemed clear. Thomas Edison and Joseph Swan independently invented useable light bulbs in the late 1870s. At the beginning of the 1880s, Edison built electricity generating stations at Pearl Street in Manhattan and Holborn in London. Things moved quickly: within a year, he was selling electricity as a commodity; a year later, the first electric motors were used to drive manufacturing machinery. Yet by 1900, less than 5 per cent of mechanical drive power in American factories was coming from electric motors. Most factories were still in the age of steam.

A steam-powered factory must have been awe-inspiring. The mechanical power came from a single massive steam engine. The engine turned a central steel drive shaft that ran along the length of the factory; sometimes it would run outside and into a second building. Subsidiary shafts, connected via belts and gears, drove hammers and punches and presses and looms. Sometimes the belts would transfer power vertically through a hole in the ceiling to a second floor, or a third. Expensive 'belt towers' enclosed them to prevent fires from spreading through the gaps. Everything was continually lubricated by thousands of drip oilers.

Steam engines rarely stopped. If a single machine in the factory needed to run, the coal fires needed to be fed. The cogs whirred and the shafts span and the belts churned up the grease and the dust, and there was always the risk that a worker might snag a sleeve or bootlace and be dragged into the relentless, all-embracing machine.

Some factory owners experimented by replacing a steam engine with an electric motor, drawing clean and modern power from a nearby generating station. After such a big investment, they tended to be disappointed with the cost savings. And it wasn't just that people didn't want to scrap their old steam engines. They kept installing more. Until around 1910, plenty of entrepreneurs looked at the old steam-engine system and the new electrical drive system, and opted for good old-fashioned steam. Why?

The answer was that to take advantage of electricity, factory owners had to think in a very different way. They could, of course, use an electric motor in the same way as they used steam engines. It would slot right in to their old systems. But electric motors could do much more.

Electricity allowed power to be delivered exactly where and when it was needed. Small steam engines were hopelessly inefficient, but small electric motors worked just fine. So a factory could contain several smaller motors, each driving a small drive shaft – or, as the technology developed, every work bench would have its own machine tool with its own little electric motor. Power wasn't transmitted through a single, massive spinning drive shaft but through wires.

A factory powered by steam needed to be sturdy enough to carry huge steel drive shafts. A factory powered by electricity could be light and airy. Steam-powered factories had to be arranged on the logic of the drive shaft; electricity meant you could arrange factories on the logic of a production line. Old factories were dark and dense, packed around the shafts. New factories could spread out, with wings and windows to bring natural light and air. In the old factories, the steam engine set the pace. In the new factories, workers could set the pace.

Factories could be cleaner and safer. They could be more efficient, because machines only needed to run when they were being used.

But – and this was a big 'but' – you couldn't get these results simply by ripping out the steam engine and replacing it with an electric motor. You needed to change everything, including the architecture and the production process. And because workers would have more autonomy and flexibility, you even had to change the way they were recruited, trained and paid.

Factory owners hesitated, for understandable reasons. Of course they didn't want to scrap their existing capital. But maybe, too, they simply struggled to think through the implications of a world in which everything needed to adapt to the new technology.

In the end, change happened. It was unavoidable. Partly, of course, it was simply that mains electricity was becoming cheaper and more reliable.

But American manufacturing was also shaped by unexpected forces. One of them was the resurgence in the late 1910s and the 1920s of an invention we've already encountered: the passport. Thanks to a series of new laws that limited immigration from a war-torn Europe, average wages soared. Hiring workers became more about quality, and less about quantity. Trained workers could use the autonomy that electricity gave them. The passport helped to kickstart the dynamo.

And as more factory owners figured out how to make the most of electric motors, new ideas about manufacturing spread. Come the 1920s, productivity in American manufacturing soared in a way never seen before or since. You would think that kind of leap forward must be explained by a new technology. But no. Paul David, an economic historian, gives much of the credit to the fact that manufacturers finally figured out how to use technology that was nearly half a century old. They had to change an entire system: their buildings, their logistics and their personnel policies were all transformed to take advantage of the electric motor. And it took about fifty years.

Which puts Solow's quip in a new light. By 2000, about half a century after the first computer program, productivity was picking up a bit. Two economists, Erik Brynjolfsson and Lorin Hitt, published research showing that many companies had invested in computers for little or no reward, but others had reaped big benefits. What explained the difference? Why did computers help some companies, but not others? It was a puzzle.

Brynjolfsson and Hitt revealed their solution: what mattered was whether the companies had also been willing to reorganise as they installed the new computers, taking advantage of their potential. That often meant decentralising, outsourcing, streamlining supply chains and offering more choice to customers. You couldn't just take your old processes and add better computers any more than you could take your old steam-powered factory and add electricity. You needed to do things differently; you needed to change the whole system.

The Web is much younger even than the computer, of course. It was barely a decade old when the dotcom bubble burst. When the electric dynamo was as old as the Web is now, factory owners were still attached to steam. The really big changes were only just appearing on the horizon.

The thing about a revolutionary technology is that it changes everything – that's why we call it revolutionary. And changing everything takes time, and imagination and courage – and sometimes just a lot of hard work.

17

The Shipping Container

The most obvious feature of the global economy is exactly that: it's global. Toys from China, copper from Chile, T-shirts from Bangladesh, wine from New Zealand, coffee from Ethiopia and tomatoes from Spain. Like it or not, globalisation is a fundamental feature of the modern economy.

Statistics back this up. In the early 1960s, world merchandise trade was less than 20 per cent of world GDP. Now, it's around 50 per cent. Not everyone is happy about this; there's probably no other issue about which the anxieties of ordinary people are so in conflict with the near-unanimous approval of economists. And so controversy rages.

The arguments over trade tend to frame globalisation as a policy – maybe even an ideology, fuelled by acronymic trade deals such as TRIPS, TTIP and the TFP. But perhaps the biggest enabler of globalisation isn't a free trade agreement, but a simple invention: a corrugated steel box, 8 feet wide, 8.5 feet high and 40 feet long. A shipping container.

To understand why the shipping container has been so important, consider how a typical trade journey looked before it was invented. In 1954, an unremarkable cargo ship, the S.S. *Warrior*, carried merchandise from Brooklyn in New York to Bremerhaven in Germany. On that trip, just over 5000 tonnes of cargo – from food to household goods, letters to vehicles – was being carried as 194,582 separate items in 1156 different shipments. The record-keeping alone, keeping track of all those consignments as they moved around the dockside warehouses, was a nightmare.

But the real challenge was physically loading ships like the *Warrior*. The longshoremen who did the job would pile barrels of olives and boxes of soap onto a wooden pallet on the dock. The pallet would be hoisted in a sling and deposited in the hold of a ship, from where more longshoremen would carry or cart each item into a snug corner of the ship, poking and pulling at the merchandise with steel hooks until it settled into place against the curves and bulkheads of the hold, skilfully packing the cargo so that it would not shift on the high seas. There were cranes and forklifts available, but in the end much of the merchandise, from bags of sugar heavier than a man to metal bars the weight of a small car, needed to be shifted with muscle power.

This was far more dangerous work than manufacturing or even construction. In a large port, someone would be killed every few weeks. In 1950, New York averaged half a dozen serious incidents every day – and New York's port was one of the safer ones.

Researchers studying the S.S. *Warrior*'s trip to Bremerhaven concluded that the ship had taken ten days to load and unload, as much time as it had done for the vessel to cross the Atlantic Ocean. In total, the cargo cost around $420 a tonne to move, in today's money. Given typical delays in sorting and distributing the cargo by land, the whole journey might take three months.

Sixty years ago, then, shipping goods internationally was costly, chancy and immensely time-consuming. Surely there was a better way. Indeed there was: put all the cargo into big standard boxes, and move the boxes.

But inventing the box was the easy bit – the shipping container had already been tried in various forms for decades, without catching on. The real challenge was overcoming the social obstacles. To begin with, the trucking companies, shipping companies and ports couldn't agree on a standard. Some wanted large containers; others wanted smaller or shorter versions, perhaps because they specialised in heavy goods such as canned pineapple, or moved them by truck on narrow mountain roads.

Then there were the powerful dock-workers' unions, who resisted the idea. You might think they'd have welcomed shipping containers, as they'd make the job of loading ships safer – but they also meant there'd be fewer jobs to go around.

Stodgy US regulators also preferred the status quo. The freight sector was tightly bound with red tape, with separate sets of regulations determining how much shipping and trucking companies could charge. Why not simply let companies charge whatever the market would bear – or even allow shipping and trucking companies to merge, and put together an integrated service? Perhaps the bureaucrats, too, were simply keen to preserve their jobs; such bold ideas would have left them with nothing to do.

The man who navigated this maze of hazards – who can fairly be described as the inventor of the modern shipping container system – was an American, Malcom (born Malcolm) McLean. McLean didn't know anything about shipping. But he was a trucking entrepreneur – he knew plenty about trucks, a lot about the system and all there was to know about saving money. Tales of McLean's penny pinching abound. As a young trucker, the story goes, he was so poor that he couldn't pay the toll at a bridge; he left his wrench as a deposit at the toll booth and redeemed the debt on his return journey, having sold his cargo. Even when McLean was in charge of a large organisation, he instructed his employees to keep long-distance phone calls briefer than three minutes, to save money.

But McLean's biographer Marc Levinson, who wrote the definitive history of the shipping container, argues that such tales don't capture the ambition or the boldness of the man. McLean saw the potential of a shipping container that would fit neatly onto a flatbed truck, but he wasn't the first person to propose such an approach. What made McLean different was his political savvy and his daring – attributes that were essential in bringing about a massive change to the global freight system.

For example: in what Levinson describes as 'an unprecedented piece of financial and legal engineering', McLean managed to gain control of both a shipping company and a trucking company at the same time. This was, of course, a great help in introducing containers that were compatible with both ships and trucks. McLean was also able to make progress with more straightforward pieces of entrepreneurship: for example, when dock workers were threatening to strike and shut down ports on the US east coast in 1956, McLean decided that this was the perfect time to refit old ships to new container specifications. He wasn't averse to plunging into debt to make the investments necessary. By 1959 he was widely suspected of being close to bankruptcy, always a risk of an ambitious debt-funded expansion. But he pulled through.*

---

* Eventually one of these gambles failed to pay off; in 1986, when McLean was seventy-two years old, his vast business was driven into bankruptcy. McLean had invested heavily in highly fuel-efficient ships, but the oil price fell sharply, making that costly investment unprofitable. Five years later, McLean was back in business again. He clearly loved his calling as an entrepreneur.

McLean was also a shrewd political operative. For example, when New York's Port Authority was trying to expand its influence in the 1950s, he pointed out that the New Jersey side of the harbour was underused and would be the perfect place for a purpose-built container shipping facility. As a result, he was able to get a large foothold in New York with both political and financial cover from the Port Authority.

But perhaps the most striking coup took place in the late 1960s, when Malcom McLean sold the idea of container shipping to perhaps the world's most powerful customer: the US military. Faced with an unholy logistical nightmare in trying to ship equipment to Vietnam, the military turned to McLean and his container ships to sort things out. Containers work much better when they're part of an integrated logistical system, and the US military was perfectly placed to adopt that system wholesale. Even better, McLean realised that on the way back from Vietnam, his empty container ships could collect payloads from the world's fastest growing economy, Japan. And so the trans-Pacific trading relationship began in earnest.

A modern shipping port today would be unrecognisable to a hardworking longshoreman of the 1950s. Even a modest container ship might carry twenty times as much cargo as the S.S. *Warrior* did, yet disgorge its cargo in hours rather than days. Gigantic cranes, weighing 1000 tonnes apiece, will lock onto containers that weigh upwards of 30 tonnes and swing them up and over onto a waiting transporter. The colossal ballet of engineering is choreographed by computers, which track every container as it moves through a global logistical system. The refrigerated containers are put in a hull section with power and temperature monitors. The heavier containers are placed at the bottom to keep the ship's centre of gravity low; the entire process is designed and scheduled to keep the ship balanced. And after the crane has released one container onto a waiting transporter, it will grasp another before swinging back over the ship, which is being emptied and refilled simultaneously.

Not everywhere enjoys the benefits of the containerisation revolution: many ports in poorer countries still look like New York in the 1950s. Sub-Saharan Africa, in particular, remains largely cut off from the world economy because of poor infrastructure. Without the ability to plug into the world's container shipping system, Africa becomes a costly place to do business with.

But for an ever growing number of destinations, goods can now be shipped reliably, swiftly and cheaply: rather than the $420 that a customer would have paid to get the S.S. *Warrior* to ship a tonne of goods across the Atlantic in 1954, you might now pay less than $50 a tonne. As a result, manufacturers are less and less interested in positioning their factories close to their customers – or even their suppliers. What matters instead is finding a location where the workforce, the regulations, the tax regime and the going wage all help make production as efficient as possible. Workers in China enjoy new opportunities; in developed countries they experience new threats to their jobs; and governments anywhere feel that they're competing with governments everywhere to attract business investment. On top of it all, in a sense, is the consumer, who enjoys the greatest possible range of the cheapest possible products – toys, phones, clothes, anything. And underpinning it all is a system: the system that Malcom McLean developed and guided through its early years.

The world is a very big place, but these days the economists who study international trade often assume that transport costs are zero. It keeps the mathematics simpler, they say – and thanks to the shipping container, it's nearly true.

18

The Barcode

There are two ways to tell this story.

One of them describes that classic flash of inventive insight. In 1948, Joseph Woodland, a graduate student at the Drexel Institute in Philadelphia, was pondering a challenge that had been thrown down by a local Philly retailer: was there any way to speed up the process of checking out in his stores by automating the tedious process of recording the transaction?

Woodland was a smart young man. He'd worked on the Manhattan Project during the war. At the other end of the spectrum, as an undergraduate he'd designed an improved system for playing elevator music. He had planned to launch it commercially, but had been warned away by his father, who was convinced that mobsters controlled the elevator music racket.

So Woodland had returned to his studies at the Drexel Institute, and now he was stumped by the transaction problem. On a visit to his grandparents down in Miami he sat on the beach, pondered and idly combed his fingers in a circle, letting the sand slide between his fingertips. But then, as he looked down at the ridges and furrows, a thought struck him. Just like Morse code used dots and dashes to convey a message, he could use thin lines and thick lines to encode information. A zebra-striped bullseye could describe a product and its price in a code that a machine might be able to read.

The idea was workable but with the technology of the time it was costly. But as computers advanced and lasers were invented, it became more realistic. The striped-scan system was independently rediscovered and refined several times over the years. In the 1950s, an engineer, David Collins, put thin and thick lines on railway cars so that they could be read automatically by a trackside scanner. In the early 1970s, an IBM engineer, George Laurer, figured out that a rectangle would be more compact than Woodland's bullseye, and developed a system that used lasers and computers that were so quick they could process labelled beanbags being hurled over the scanner system. Joseph Woodland's seaside doodles had become a technological reality.

But there's a second way to tell the story. It's just as important – but it's incredibly boring.

In September 1969, members of the Administrative Systems Committee of the Grocery Manufacturers of America (GMA) met with their opposite numbers from the National Association of Food Chains (NAFC). The location: a motel. The Carousel Inn in Cincinnati. It wasn't terribly good. The topic? Ah, the topic. The topic was whether the food producers of the GMA could come to an agreement with the food retailers of the NAFC about an inter-industry product code.

The GMA wanted an eleven-digit code, which would encompass various labelling schemes it was already using. The NAFC wanted a shorter, seven-digit code, which could be read by simpler and cheaper systems at the checkout.

The GMA and the NAFC couldn't agree, and the meeting broke up in frustration. Years of careful diplomacy – and innumerable committees, subcommittees and ad hoc committees – were required before, finally, the US grocery industry agreed upon a standard for the Universal Product Code, or UPC.

Both stories came to fruition in June 1974 at the checkout counter of Marsh's Supermarket in the town of Troy, Ohio, when a 31-year-old checkout assistant named Sharon Buchanan scanned a ten-pack of fifty sticks of Wrigley's Juicy Fruit chewing gum across a laser scanner, automatically registering the price of 67 cents. The gum was sold. The barcode had been born.

We tend to think of the barcode as a simple piece of cost-cutting technology: it helps supermarkets do their business more efficiently, and so it helps us to enjoy lower prices. But like the shipping container, the barcode doesn't work unless it's integrated into a system. And like the container shipping system, the barcode system does more than lower costs. It solves problems for some players and creates headaches for others.

That is why the second way of telling the story is as important as the first – because the barcode changes the balance of power in the grocery industry. It's why all those committee meetings were necessary; it's why the food retailing industry was able to agree, in the end, only when the technical geeks on the committees were replaced by their bosses' bosses, the chief executives. The stakes were that high.

Part of the difficulty was getting everyone to move forward on a system which didn't really work without a critical mass of adopters. It was expensive to install scanners. It was expensive to redesign packaging with barcodes – bear in mind that Miller's Beers were still printing labels for their bottles on a 1908 printing press. The retailers didn't want to install scanners until the manufacturers had put barcodes on their products; the manufacturers didn't want to put barcodes on their products until the retailers had installed enough scanners.

But it also became apparent over time that the barcode was changing the tilt of the playing field in favour of a certain kind of retailer. For a small, family-run convenience store, the barcode scanner was an expensive solution to problems it didn't really have. But big-box supermarkets could spread the cost of the scanners across many more sales. They valued shorter lines at the checkout. They needed to keep track of inventory. With a manual checkout, a shop assistant might charge a customer for a product, then slip the cash into her pocket without registering the sale. With a barcode and scanner system, such behaviour would be pretty conspicuous. And in the 1970s, a time of high inflation in America, barcodes let supermarkets change the price of products by sticking a new price tag on the shelf rather than on each item.

It's hardly surprising that as the barcode spread across retailing in the 1970s and 1980s, large retailers also expanded. The scanner data underpinned customer databases and loyalty cards. By tracking and automating inventory, it made just-in-time deliveries more attractive, and lowered the cost of having a wide variety of products. Shops in general, and supermarkets in particular, started to generalise – selling flowers, clothes and electronic products. Running a huge, diversified, logistically complex operation was all so much easier in the world of the barcode.

Perhaps the ultimate expression of that fact came in 1988 – when the discount department store Wal-Mart decided to start selling food. It is now the largest grocery chain in America – and by far the largest general retailer on the planet, about as large as its five closest rivals combined. Wal-Mart was an early adopter of the barcode and has continued to invest in cutting-edge computer-driven logistics and inventory management.

Wal-Mart is now a major gateway between Chinese manufacturers and American consumers. Its embrace of technology helped it grow to a vast scale, and its vast scale means it can send buyers to China and commission cheap products in bulk. From a Chinese manufacturer's perspective, you can justify setting up an entire production line for just one customer – as long as that customer is Wal-Mart.

Geeks rightly celebrate the moment of inspiration as Joseph Woodland languidly pulled his fingers through the sands of Miami Beach – or the perspiration of George Laurer as he perfected the barcode as we know it. But the barcode isn't just a way to do business more efficiently; it also changes what kind of business can be efficient.

The barcode is now such a symbol of the forces of impersonal global capitalism that it has spawned its own ironic protest. Since the 1980s, people have been registering their opposition by getting themselves tattooed with a barcode. That countercultural fashion statement recognises something important. Yes, those distinctive black and white stripes are a neat little piece of engineering. But that neat little piece of engineering has changed how the world economy fits together.

19

The Cold Chain

'Crazier than a half-dozen opium-smoking frogs': that's how one observer described General Jorge Ubico. The General, who was president of Guatemala from 1931 to 1944, liked to dress up as Napoleon Bonaparte. He may even have believed himself to *be* Napoleon Bonaparte, reincarnate.

Like many twentieth-century Latin American dictators, crazy General Ubico had a cosy relationship with the United Fruit Company. It became known as *El Polpo*, the octopus, because its tentacles reached everywhere. Ubico passed a law forcing indigenous Guatemalans to work for landowners – which is to say, the United Fruit Company, which owned most of Guatemala's arable land. And United left most of it lying fallow, just in case it might need it in future. The company claimed that the land was worth next to nothing, so it shouldn't have to pay much tax on it. Ubico agreed.

But then Ubico was overthrown. An idealistic young soldier, Jacobo Árbenz, rose to power. And he called *El Polpo*'s bluff. If the land was worth so little, the state would buy it and let peasants farm it. The United Fruit Company didn't like this idea. It lobbied the US government, employing a PR agency to portray Árbenz as a dangerous communist. The CIA got involved. In 1954, Árbenz was ousted in a coup, stripped to his underwear and bundled onto a flight and peripatetic exile. His daughter killed herself. He drank himself into oblivion, and died with a bottle of whiskey in a hotel bathtub. Guatemala descended into a civil war that lasted for thirty-six years.

There's a name for poor countries with crazy dictators propped up by cynical foreign money: banana republics. Ironically, Guatemala's woes were intimately bound up with its chief export, bananas. But without the invention of yet another system, Guatemalan politics – and western diets – would look very different. That system is called the cold chain.

Long before the cold chain existed, one of the co-founders of the United Fruit Company was a man called Lorenzo Dow Baker. He started off as a sailor. In 1870, he'd just ferried some gold prospectors up the Orinoco River, and his boat sprang a leak on the way home to New England, so he docked in Jamaica for repairs. He had money in his pocket, and he liked to gamble – so he bought bananas, backing himself to get them home before they spoiled. He managed it, just, sold them for a healthy profit, and went back for more. Bananas became a delicacy in port cities like Boston and New York. Ladies ate them with a knife and fork, to avoid any embarrassing sexual connotations.

But bananas were a risky business. Their shelf life was right on the cusp of the sailing time; when they arrived, they were too ripe to survive the onward journey inland. If only you could keep them cool en route, they'd ripen more slowly, and reach a bigger market.

Bananas weren't the only foodstuff prompting interest in refrigerating ships. Two years before Baker's first journey from Jamaica, Argentina's government offered a prize to anyone who could keep its beef cold enough for long enough to export it overseas. Packing ships with ice had led to costly failures. For a century, scientists had known that you could artificially lower the temperature by compressing some gases into liquids, then letting the liquid absorb heat as it evaporated again. But commercially successful applications remained elusive. In 1876, Charles Tellier, a French engineer, fitted up a ship with a refrigeration system, packed it with meat, and sailed it to Buenos Aires as proof of concept: after 105 days at sea, the meat arrived still fit to eat.

La Liberté, an Argentinian newspaper, rejoiced: 'Hurray, a thousand times for the revolutions of science and capital'; Argentinian beef exports could begin. By 1902, there were 460 refrigerated ships – or 'reefers' – plying the world's seas, carrying a million tonnes of Argentina's beef, El Polpo's bananas and much else besides.

Meanwhile, in Cincinnati, a young African-American boy was facing up to life as an orphan. He dropped out of school at the age of twelve, got a job sweeping the floor at a garage and learned how to mend cars. His name was Frederick McKinley Jones, and he grew up to be a prolific inventor. By 1938 he was working as a sound systems engineer, when his boss's friend – who, like Malcom McLean, ran a trucking business – complained about the difficulties of transporting perishable goods by land. Reefers' refrigeration units couldn't cope with the vibration of road travel, so you still had to pack your trucks with ice, and hope you'd complete the journey before the ice melted. That didn't always happen. Could the brilliant, self-taught Jones invent a solution?

He could. The result was a new company, Thermo King, and the last link in the cold chain – the global supply chain that keeps perishable goods at controlled temperatures. The cold chain revolutionised healthcare. During the Second World War, Jones's portable refrigeration units preserved drugs and blood for injured soldiers. The cold chain is how vaccines get around without going bad, at least until they reach remote parts of poor countries with unreliable power supplies – and there are new inventions on the horizon to solve that problem.

Above all, the cold chain revolutionised food. On a warm summer's day – let's say 25 degrees Celsius – fish and meat will last only a few hours; fruit will be mouldy in a few days; carrots might survive for three weeks if you're lucky. In the cold chain, fish will keep for a week, fruit for months and root vegetables for up to a year. Freeze the food, and it lasts for longer still.

Refrigeration widened our choice of food: tropical fruits like bananas could now reach anywhere. It improved our nutrition. It enabled the rise of the supermarket: if your home has no way to keep food cold, you have to make frequent trips to the market; with a fridge-freezer at home, you can do a big shop every week or two. And just as with the development of the TV dinner, simplifying the process of feeding a family transforms the labour market. Fewer shopping trips means housewives face fewer obstacles to becoming career women. As low-income countries get wealthier, fridges are among the first things people buy: in China, it took just a decade to get from a quarter of households having fridges to nearly nine in ten.

The cold chain is one of the pillars of the global trading system. As we've seen, the shipping container made long-distance commerce cheaper, quicker and more predictable. The barcode helped huge, diverse retailers keep track of complex supply chains. The diesel engine – which we'll encounter later in the book – made huge ocean-going ships amazingly efficient.

And the cold chain? The cold chain took all these other inventions and extended their reach to perishable food. Now meat, fruit and vegetables were subject to the economic logic of global specialisation and global trade. Yes, you can grow French beans in France – but perhaps you should fly them in from Uganda? Different growing conditions mean this kind of thing can make environmental sense, as well as economic. One study found it was eco-friendlier to grow tomatoes in Spain and transport them to Sweden than to grow them in Sweden. Another claimed that it emits less carbon to raise a lamb in New Zealand and ship it to England than to raise a lamb in England.

Economic logic tells us that specialisation and trade will increase the value of production in the world. Sadly, it doesn't guarantee that value will be shared fairly. Consider the state of Guatemala today. It still exports bananas – hundreds of millions of dollars' worth. It breeds and grows lots of other stuff, too: sheep, sugarcane, coffee, corn and cardamom. But it has the world's fourth-highest rate of chronic malnutrition – half its kids are stunted because they don't get enough to eat.

Economists still don't fully understand why some countries grow rich while others stay poor, but most agree on the importance of institutions – things like corruption, political stability and the rule of law. According to one recent ranking of countries' institutions, Guatemala came a lowly 110th of 138. The legacy of General Ubico, the banana-driven coup, and the civil war lives on. Cold chain technologies were designed to make bananas last longer. But banana republics, it seems, have a naturally long shelf-life.

20

Tradable Debt and the Tally Stick

Not far from my home is Oxford's Ashmolean Museum, home to art and antiquities from around the world. I often find myself slipping down the stairs to the grand basement, and because I'm an economist I bypass the café and head instead to the money gallery next to it. You can see coins from Rome, the Vikings, the Abbasid Caliphate and closer to home from medieval Oxfordshire and Somerset. But while it seems obvious that the money gallery would be full of coins, most money isn't in the form of coins at all.

As Felix Martin points out in his book *Money: The Unauthorised Biography*, we tend to misunderstand money because much of our monetary history hasn't survived in a form that could grace a museum. In fact, in 1834, the British government decided to destroy six hundred years of precious monetary artefacts. It was a decision that was to have unfortunate consequences in more ways than one.

The artefacts in question were humble sticks of willow, about 8 inches long, called Exchequer tallies. The willow was harvested along the banks of the Thames, not far from the Palace of Westminster in central London. Tallies were a way of recording debts with a system that was sublimely simple and effective. The stick would contain a record of the debt, carved into the wood. It might say, for example, '9£ 4s 4p from Fulk Basset for the farm of Wycombe'. Fulk Basset, by the way, might sound like a character from *Star Wars* but was in fact a Bishop of London in the thirteenth century. He owed his debt to King Henry III.

Now comes the elegant part. The stick would be split in half, down its length from one end to the other. The debtor would retain half, called the 'foil'. The creditor would retain the other half, called the 'stock' – even today, British bankers use the word 'stocks' to refer to debts of the British govern-ment. Because willow has a natural and distinctive grain, the two halves would match only each other.

Of course, the Treasury could simply have kept a record of these transactions in a ledger somewhere. But the tally stick system enabled something radical to occur. If you had a tally stock showing that Bishop Basset owed you five pounds, then unless you worried that Bishop Basset wasn't good for the money, the tally stock itself was worth close to five pounds in its own right. If you wanted to buy something, you might well find that the seller would be pleased to accept the tally stock as a safe and convenient form of payment.

The tally sticks became a kind of money – and a particu-larly instructive kind of money, too, because they show us clearly what money really is: it's debt; a particular kind of debt, one that can be traded freely, circulating from person to person until it is utterly separated from Bishop Basset and a farm in Wycombe. It's a spontaneous transformation from a narrow record of a debt to a much broader system of tradable debts.

We don't have a good sense of how significant this system is, because for unfortunate reasons that will become clear, we don't know how widely traded tally sticks were. But we know that similar debts were widely traded. This happened in China about a thousand years ago, when – as we'll see later – the idea of paper money itself emerged. But it's also happened more than once in living memory.

On Monday 4 May 1970, the *Irish Independent*, Ireland's leading newspaper, published a matter-of-fact notice with a straightforward title: CLOSURE OF BANKS. Every major bank in Ireland was closed and would remain closed until further notice. The banks were in dispute with their own employees, the employees had voted to strike, and it seemed likely that the whole business would drag on for weeks or even months.

You might think that such news – in what was one of the world's more advanced economies – would inspire utter panic, but the Irish remained calm. They'd been expecting trouble, so they had been stockpiling reserves of cash, but what kept the Irish economy going was something else.

The Irish wrote each other cheques. Now, at first sight this makes no sense. Cheques are paper-based instructions to transfer money from one bank account to another. But if both banks are closed, then the instruction to transfer money can't be carried out. Not until the banks open, anyway. But everyone in Ireland knew that might not happen for months.

Nevertheless, the Irish wrote each other cheques. And those cheques would circulate. Patrick would write a cheque for twenty pounds to clear his tab at the local pub. The publican might then use that cheque to pay his staff, or his suppliers. (It could either be made out to 'cash', or countersigned to transfer ownership.) Patrick's cheque would circulate around and around, a promise to pay twenty pounds that couldn't be fulfilled until the banks reopened and started clearing the backlog.

The system was fragile. It was clearly open to abuse by people who wrote cheques they knew would eventually bounce. As May dragged past, then June, then July, there was always the risk that people would lose track of their own finances, too, and start unknowingly writing cheques they couldn't afford and wouldn't be able to honour. Perhaps the biggest risk of all was that trust would start to fray, that people would simply start refusing to accept cheques as payment.

Yet the Irish kept writing each other cheques. It must have helped that so much Irish business was small and local. People knew their customers. They knew who was good for the money. Word would get around about people who cheated. And the pubs and corner shops were able to vouch for the creditworthiness of their customers, which meant that cheques could circulate.

When the dispute was resolved and the banks reopened in November, more than six months after they had closed, the Irish economy was still in one piece. The only problem: the backlog of five billion pounds' worth of cheques would take another three months to clear.

Nor is the Irish case the only one in which cheques were passed around without ever being cashed in. In the 1950s, British soldiers stationed in Hong Kong would pay their bills with cheques on accounts back in England. The local merchants would circulate the cheques, vouching for them with their own signatures, without any great hurry to cash them in. In effect, the Hong Kong cheques – like the Irish cheques, and like the tally sticks – had become a form of private money.

If money is simply tradable debt, then tally sticks and un-cashed Irish cheques weren't some weird form of quasi-money. They *were* money: they were simply money in a particularly unvarnished form. Like an engine running with the cover off, or a building with the scaffolding still up, they're the system of money with the underlying mechanism laid bare.

Of course, we still naturally think of money as those discs of metal in the Ashmolean Museum. After all it's the metal that survives, not the cheques or the tally sticks. And one thing you can never put into a display case at a museum is a system of trust and exchange – which, ultimately, is what modern money is.

Those tally sticks, by the way, met an unfortunate end. The system was finally abolished and replaced by paper ledgers in 1834 after decades of attempts to modernise. To celebrate, it was decided to burn the sticks – six centuries of irreplaceable monetary records – in a coal-fired stove in the House of Lords, rather than letting parliamentary staff take them home for firewood. Now burning a cartload or two of tally sticks in a coal-fired stove is a wonderful way to start a raging chimney fire. So it was that the House of Lords, then the House of Commons and almost the entire Palace of Westminster – a building as old as the tally stick system itself – was burned to the ground. Perhaps the patron saints of monetary history were having their revenge.

21

Billy Bookcase

Denver Thornton hates the Billy bookcase.

He runs a company called unflatpack.com. If you buy flatpack furniture from somewhere like Ikea, but you're terrified by dowels and allen keys and cryptic instruction leaflets featuring happy cartoon men, you can get someone like Mr Thornton to come to your house to build it for you.

And the Billy bookcase? It's the archetypal Ikea product. It was dreamed up in 1978 by an Ikea designer called Gillis Lundgren – he sketched it on the back of a napkin, worried that he'd forget it. Now there are 60-odd million in the world, nearly one for every hundred people. Not bad for a humble bookcase. In fact, so ubiquitous are they, Bloomberg uses them to compare purchasing power across the world. According to the Bloomberg Billy Bookcase Index – yes, that's a thing – they cost most in Egypt, just over a hundred dollars; in Slovakia you can get them for less than forty.

Every three seconds, another Billy bookcase rolls off the production line of the Gyllensvaans Möbler factory in Kättilstorp, a tiny village in southern Sweden. The factory's couple of hundred employees never actually touch a bookshelf – their job is to tend to the machines, imported from Germany and Japan, which work twenty-four hours to cut and glue and drill and pack the various component parts of the Billy bookcase. In goes particle board by the truck load, 600 tons a day; out come ready-boxed products, stacked six-by-three on pallets and ready for the trucks.

In the reception at the Gyllensvaans Möbler factory, in a frame on the wall, is a typewritten letter – the company's very first furniture-making order from Ikea. The date of the letter: 1952.

Ikea was not, back then, the global behemoth it is today, with stores in dozens of countries and turnover in the tens of billions. Its founder, Ingvar Kamprad, was just seventeen when he started the business with a small gift of cash from his dad, a reward for trying hard at school despite dyslexia. By 1952, aged twenty-six, young Ingvar had already got a hundred-page furniture catalogue, but hadn't yet hit on the idea of flat-packing. That came a few years later, as he and his company's fourth employee – one Gillis Lundgren – were packing a car with furniture for a catalogue photoshoot. 'This table takes up too much darn space,' Gillis said. 'We should unscrew the legs.'

It was a light–bulb moment. Kamprad was already obsessed with cutting prices – so obsessed that some manufacturers had started to boycott him. And one way to keep prices low is to sell furniture in bits, rather than paying labourers to assemble it. In that sense, it may seem perverse to get a Denver Thornton in to construct your Billy bookcase – it's a bit like buying ingredients at a supermarket, and hiring a private chef to cook your dinner.

And that might be true if outsourcing the labour to the customer was the only thing that made flatpacks cheaper. But even bigger savings come from precisely the problem that inspired Gillis Lundgren: transport. In 2010, for example, Ikea rethought the design of its Ektorp sofa. It made the armrests detachable. That helped halve the size of the packaging, which halved the number of trucks needed to get the sofas from factory to warehouse, and from warehouse to store. And that lopped a seventh off the price – more than enough to cover Mr Thornton's labour for screwing on the armrests.

It's not just furniture that benefits from a constant questioning of product design. Consider another Ikea icon: the Bang mug. You've probably had a drink from one – with yearly sales reaching twenty-five million, there are plenty of them knocking around. Its design is distinctive – wide at the top, tapering downwards; a small handle, right by the rim – and it's not motivated purely by aesthetics. Ikea changed the height of the mug when it realised that it could make slightly better use of the space in its supplier's kiln in Romania. And by tweaking the handle design, Ikea made them stack more compactly – more than doubling the number you could fit on a pallet, and more than halving the cost of getting them from the kiln in Romania to the shelves in the store.

It's been a similar story with the Billy bookcase. It doesn't look like it's changed much since it was designed in the late 1970s, but it does cost 30 per cent less. That's partly due to constant, tiny tweaks in both product and production method. But it's also thanks to sheer economies of scale – the more of something you can commit to making, the cheaper you can get it made. Look at Gyllensvaans Möbler: compared to the 1980s, it's making thirty-seven times as many bookcases, yet its number of employees has only doubled. Of course, that's thanks to all those German and Japanese robots. Yet a business needs confidence to sink so much money into machinery, especially when it has no other client: pretty much all Gyllensvaans Möbler does is make bookcases for Ikea.

Or consider, again, the Bang mug. Initially, Ikea asked a supplier to price up a million units in the first year. Then it said: what if we commit to five million a year for three years? That cut the cost by a tenth. Not much, perhaps, but every penny counts. Just ask the famously penny-pinching Ingvar Kamprad: in a rare interview to mark his ninetieth birthday, Kamprad claimed to be wearing clothes he'd bought at a flea market. He is said to fly economy and drive an old Volvo. This frugality may help to explain why he's the world's eighth-richest man – although the four decades he spent living in Switzerland to avoid Swedish taxes may also have something to do with it.

Still, penny-pinching isn't all it takes to succeed. Anyone can make shoddy, ugly goods by cutting corners. And anyone can make elegant, sturdy products by throwing money at them. To get as rich as Kamprad has, you have to make stuff that's both cheap and acceptably good.

And that's what seems to explain the enduring popularity of the Billy bookcase. 'Simple, practical and timeless' is how Gillis Lundgren once described the designs he hoped to create, and Billy is surprisingly well accepted by the type of people you might expect to be sniffy about mass-produced MDF. Sophie Donelson, who edits the interiors magazine *House Beautiful*, told AdWeek that Billy is 'unfussy' and 'unfettered', and 'modern without trying too hard'.

Furniture designer Matthew Hilton praises an interesting quality: anonymity. Interiors creative Mat Sanders agrees, declaring that Ikea is 'a great place for basic you can really dress up to make feel high-end'. Billy is a bare-bones, functional bookshelf if that's all you want from it, or it's a blank canvas for creativity: on ikeahackers.net you'll see it repurposed as everything from a wine rack to a room divider to a baby-changing station.

But business and supply chain nerds don't admire the Billy bookcase for its modernity or flexibility. They admire it – and Ikea in general – for relentlessly finding ways to cut costs and prices without reducing the quality of the product. That is why Billy is a symbol of how innovation in the modern economy isn't just about snazzy new technologies, but about boringly efficient systems. The Billy bookcase isn't innovative in the way that the iPhone is innovative. The innovations are about working within the limits of production and logistics – finding tiny ways to shave more off the cost, all while producing something that looks inoffensive and does the job.

And that annoys handyman Denver Thornton. 'It's just so easy and monotonous,' he says. 'I prefer a challenge.'

22

The Elevator

Here's a little puzzle.

One day, on her way to work, a woman decides that she's going to take a mass-transit system instead of her usual method. Just before she gets on board, she looks at an app on her phone that gives her position with the exact latitude and longitude. The journey is smooth and perfectly satisfactory, despite frequent stops, and when the woman disembarks she checks her phone again. Her latitude and longitude haven't changed at all. What's going on?

The answer: this lady works in a tall office building, and rather than taking the stairs, she's taken the lift. We don't tend to think of elevators as mass-transportation systems, but they are: they move hundreds of millions of people every day, and China alone is installing 700,000 elevators a year.

The tallest building in the world, the Burj Khalifa in Dubai, has more than 300,000 square metres of floor space; the brilliantly engineered Sears Tower in Chicago has more than 400,000. Imagine such skyscrapers sliced into fifty or sixty low-rise chunks, then surrounding each chunk with a car park and connecting all the car parks together with roads, and you'd have an office park the size of a small town. The fact that so many people can work together in huge buildings on compact sites is possible only because of the elevator.

Or, perhaps we should say – because of the *safety* elevator. Elevators themselves have existed for a long time, typically using the very simple principle of a rope and a pulley. Archimedes is said to have built one in ancient Greece. In 1743, at the Palace of Versailles, Louis XV used one to clandestinely visit his mistress – or, alternatively, so that his mistress could clandestinely visit him. The power for King Louis's secret love-lift was supplied by a chap in a hollow section of wall, standing ready to haul on a rope when required. Other elevators – in Hungary, in China, in Egypt – were powered by draft animals. Steam power went further: Matthew Boulton and James Watt, two giants of Britain's industrial revolution, produced steam engines that ran muscular industrial lifts, hauling coal up from the mines. But while these elevators all worked well enough, you wouldn't want to use them to lift *people* to any serious height – because, inevitably, something could go wrong. The elevator could plunge down through the shaft, loose ends of the rope flapping in the darkness, passengers screaming into oblivion. Most people can walk up five flights of stairs if they must; nobody in their right mind would want to take an *elevator* to such a deadly height.

So what mattered was making a lift that was not only safe, but demonstrably and consistently safe. Such responsibility fell to a man named Elisha Otis. At the 1853 World's Fair in New York, Otis climbed onto a platform, which was then hoisted high above a crowd of onlookers, nervy with anticipation. The entire contraption looked a little like an executioner's scaffold. Behind Otis stood a man with an axe, which can only have added to the sense that a spectacular death was about to occur. The axeman swung down onto the rope, the crowd gasped, Otis's platform shuddered – but did not plunge. 'All safe, gentlemen, all safe!' boomed Otis. The city landscape was about to be turned on its head by the man who had invented not the elevator, but the elevator brake.

'Turned on its head' is right – because the new safe elevators transformed the position of the highest status areas in the building. When the highest reaches of a six- or seven-storey building were at the end of an arduous climb, they used to be the servants' quarters, the attic for mad aunts, or the garret for struggling artists. After the invention of the elevator, the attic became the loft apartment. The garret became the penthouse.

The elevator is best understood as part of a broader system of urban design. Without the air conditioner, modern glass skyscrapers would be uninhabitable; without either steel or reinforced concrete, they would be unbuildable; and without the elevator, they would be inaccessible.

Another crucial element of that system was mass public transportation: the subways and other urban transit systems that could bring large numbers of people into dense urban cores. In the quintessential high-rise centre, Manhattan, elevators and the subway are symbiotic. The density that the skyscrapers provide makes it easier to run a subway system efficiently; without the subway system, nobody would be able to get to the skyscrapers.

The result is a surprisingly green urban environment: more than 80 per cent of Manhattanites travel to work on the subway, or by bike or on foot, ten times the rate for America as a whole. A similar story can be told for high-rise cities across the planet from Singapore to Sydney. They tend to be very desirable places to live in – as witnessed by people's willingness to pay expensive rents to do so. They're creative, as measured by the output of patents and a high rate of start-ups. They're rich, as measured by economic output per person. And relative to rural and suburban areas, they are environmental utopias, with low rates of energy use per person and low consumption of petrol. This minor miracle – wealth, creativity and vitality with a modest environmental footprint – would be impossible without the elevator.

Yet the elevator seems unfairly under-rated. We hold it to a higher standard than other forms of transport. We're pleased if we have to wait only a couple of minutes for a bus or a train, but grumble if we have to wait twenty seconds for an elevator. Many people are nervous of elevators, yet they are safe – at least ten times safer than escalators. Frankly, the elevator is a faithful servant that is too often ignored. Perhaps this is because using one feels almost like being teleported: the doors close, there is a shift in the feeling of gravity, the doors open again and you're somewhere else. There is so little sense of place, that without signs and LED displays, we wouldn't have a clue which floor we were emerging into.

While we take the lift for granted, it continues to evolve. The challenges of ever-taller skyscrapers are being met by super-light elevator ropes, and by computer controllers that will allow two lifts to shuttle up and down a single shaft independently, one above the other. But often the older, simpler ideas still work: for example, making the wait for an elevator pass more quickly by putting full length mirrors in the elevator lobby. And the elevator is naturally energy efficient because elevator cars have counterweights.

There's always room for improvement, of course. The Empire State Building – still the most iconic skyscraper in the world – was recently retrofitted in a 500-million-dollar effort to reduce the building's carbon emissions. The retrofit included elevators with regenerative brakes, so that when a full car comes down or an empty car heads up, the elevator supplies power back to the building.

But the truth is that the Empire State Building was always energy efficient by the simple virtue of being a densely packed vertical structure next to a subway station. One of the organisations that designed the building's retrofit is the visionary environmental organisation, the Rocky Mountain Institute (RMI), whose super-efficient, environmentally sustainable headquarters doubling as a showcase home for founder Amory Lovins, was built high in the Rockies, 180 miles away from the nearest public transit system. The RMI has expanded, and staff take pains to use energy-saving technology to get to meetings – electric cars, buses and teleconferencing. And the RMI is a showcase for environmentally efficient design ideas – including high-tech coatings on the windows, krypton-filled triple-glazing, a water-reuse system and energy-saving heat exchangers.

But the elevator requires no pains to be taken at all. It's one of the most environmentally friendly technologies and it is on display in buildings all around us. It's a green mode of transport that moves billions of people every year – and yet is so overlooked that it can hide in plain sight as the answer to a lateral thinking puzzle.

IV

IDEAS ABOUT IDEAS

Some of the most powerful inventions are those that allow other inventions to flourish. The barcode, the cold chain and the shipping container combined to unleash the forces of globalisation. The elevator was vastly more useful when deployed alongside steel and concrete, the subway and the air conditioner.

But nowhere is this more true than when someone develops an idea that allows other ideas to emerge. Thomas Edison arguably did just that, inventing a process for inventing things, bringing together at Menlo Park the resources needed to tinker and experiment on an industrial scale. Here's a description from 1876:

On the ground-floor, as you enter, is a little front-office, from which a small library is partitioned off. Next is a large square room with glass cases filled with models of his inventions. In the rear of this is the machine-shop, completely equipped, and run with a ten-horse-power engine. The upper story occupies the length and breadth of the building, 100 × 25 feet, is lighted by windows on every side, and is occupied as a laboratory. The walls are covered with shelves full of bottles containing all sorts of chemicals. Scattered through the room are tables covered with electrical instruments, telephones, phonographs, microscopes, spectroscopes, etc. In the centre of the room is a rack full of galvanic batteries.

Equipped with his 'invention factory', Edison reckoned that he would develop 'a minor invention every ten days and a big thing every six months or so'. It's hard to quibble with the results – Edison's name appears repeatedly in these pages.

But even Edison's invention of the invention factory itself arguably pales beside some of the other 'meta-ideas' that have been developed – ideas about how ideas should be protected, how ideas should be commercialised, and how ideas should be kept secret. And the oldest idea about ideas is almost as old as the plough itself.

23

Cuneiform

People used to believe that writing had come from the gods. The Greeks believed that Prometheus had given it to mankind as a gift. The Egyptians also thought that literacy was divine, a benefaction from baboon-faced Thoth, the god of Knowledge. Mesopotamians thought that the goddess Inanna had stolen it for them from Enki, the god of Wisdom – although Enki wasn't so wise that he hadn't drunk himself insensible.

Scholars no longer embrace the 'baboon-faced Thoth' theory of literacy. But why ancient civilisations developed writing was a mystery for a long time. Was it for religious or artistic reasons? To send messages to distant armies? The mystery deepened in 1929, when a German archaeologist named Julius Jordan unearthed a vast library of clay tablets that were five thousand years old. They were far older than the samples of writing that had been found in China, Egypt and Mesoamerica, and they were written in an abstract script that became known as 'cuneiform'.

The tablets came from Uruk, a Mesopotamian settlement on the banks of the Euphrates in what is now Iraq. Uruk was small by today's standards – more like a large village, with a few thousand inhabitants. But by the standards of five thousand years ago, Uruk was huge – one of the world's first true cities.

'He built the town wall of "Uruk", city of sheepfolds', proclaims the *Epic of Gilgamesh*, one of the earliest works of literature. 'Look at its wall with its frieze like bronze! Gaze at its bastions, which none can equal!'

But this great city had produced writing that no modern scholar could decipher. What did it say?

Uruk posed another puzzle for archaeologists – although it seemed unrelated. The ruins of Uruk and other Mesopotamian cities were littered with little clay objects, some conical, some spherical, some cylindrical. One archaeologist quipped that they looked like suppositories. Julius Jordan himself was a little more perceptive. They were shaped, he wrote in his journal, 'like the commodities of daily life: jars, loaves, and animals', although they were stylised and standardised.

But what were they for? Were they baubles? Toys for children? Pieces for boardgames? They were the right size for that, at least. Nobody could work it out.

Nobody, that is, until a French archaeologist named Denise Schmandt-Besserat. In the 1970s she catalogued similar pieces found across the region, from Turkey to Pakistan. Some of them were nine thousand years old. Schmandt-Besserat believed that the tokens had a simple purpose: correspondence counting. The tokens that were shaped like loaves could be used to count loaves. The ones shaped like jars could be used to count jars. Correspondence counting is easy: you don't need to know how to count, you just need to look at two quantities and verify that they are the same.

Correspondence counting is older even than Uruk – the Ishango Bone, found near one of the sources of the Nile in the Democratic Republic of Congo, seems to use matched tally marks on the thigh bone of a baboon for correspondence counting. It's twenty thousand years old.

But the Uruk tokens took things further, because they were used to keep track of counting lots of different quantities, and could be used both to add and to subtract. Uruk, remember, was a great city in its day. In such a city, one cannot live hand to mouth. People were starting to specialise. There was a priesthood and there were craftsmen. Food needed to be gathered from the surrounding countryside. An urban economy required trading and planning and even taxation. So picture the world's first accountants, sitting at the door of the temple storehouse, using the little loaf tokens to count as the sacks of grain arrive and leave.

Denise Schmandt-Besserat pointed out something else – something rather revolutionary. Those abstract marks on the cuneiform tablets? They matched the tokens. Everyone else had missed the resemblance because the writing didn't seem to be a picture of anything; it seemed abstract.

But Schmandt-Besserat realised what had happened. The tablets had been used to record the back-and-forth of the tokens, which themselves were recording the back-and-forth of the sheep, the grain and the jars of honey. In fact, it may be that the first such tablets were impressions of the tokens themselves, the result of pressing the hard clay baubles into the soft clay tablet.

Then those ancient accountants realised it might be simpler to make the marks with a stylus. So cuneiform writing was a stylised picture of an impression of a token representing a commodity. No wonder nobody had made the connection before Schmandt-Besserat. And so she solved both problems at once. Those clay tablets, adorned with the world's first abstract writing? They weren't being used for poetry, or to send messages to far-off lands. They were used to create the world's first accounts.

The world's first written contracts, too – since there is just a small leap between a record of what has been paid, and a record of a future obligation to pay. The combination of the tokens and the clay cuneiform writing led to a brilliant verification device: a hollow clay ball called a bulla. On the outside of the bulla, the parties to a contract could write down the details of the obligation – including the resources that were to be paid. On the inside of the bulla would be the tokens representing the deal. The writing on the outside and the tokens inside the clay ball verified each other.

We don't know who the parties to such agreements might have been – whether they were religious tithes to the temple, taxes, or private debts, is unclear. But such records were the purchase orders and the receipts that made life in a complex city society possible.

This is a big deal. Many financial transactions are based on explicit written contracts – including those described elsewhere in this book, such as insurance, bank accounts, publicly traded shares, index funds and paper money itself. Written contracts are the lifeblood of modern economic activity – and the bullas of Mesopotamia are the very first archaeological evidence that written contracts existed.

Uruk's accountants provided us with another innovation, too. At first, the system for recording five sheep would simply require five separate sheep impressions. But that was cumbersome. A superior system involved using an abstract symbol for different numbers – five strokes for five, a circle for ten, two circles and three strokes for twenty-three. The numbers were always used to refer to a quantity *of* something: there was no 'ten' – only ten sheep. But the numerical system was powerful enough to express large quantities – hundreds, and even thousands. One demand for war reparations, 4400 years old, is for 4.5 trillion litres of barley grain, or 8.64 million 'guru'. It was an unpayable bill – six hundred times the US's annual production of barley today. But it was an impressively big number. It was also the world's first written evidence of compound interest. But perhaps that is a tale for another time.

In all it is quite a set of achievements. The citizens of Uruk faced a huge problem, a problem that is fundamental to any modern economy – the problem of dealing with a web of obligations and long-range plans between people who did not know each other well, who might perhaps never even meet. Solving that problem meant producing a string of brilliant innovations: not only the first accounts and the first contracts, but the first mathematics and even the first writing.

Writing wasn't a gift from Prometheus or Thoth. It was a tool that was developed for a very clear reason: to run an economy.

24

Public Key Cryptography

Two graduate students stood silently next to a lectern, listening as their professor presented their work to a conference. This wasn't the done thing: usually, the students themselves would get to bask in the glory. And they'd wanted to, just a couple of days previously. But their families talked them out of it. It wasn't worth the risk.

A few weeks earlier, the Stanford researchers had received an unsettling letter from a shadowy agency of the United States government. If they publicly discussed their findings, the letter said, that would be deemed legally equivalent to exporting nuclear arms to a hostile foreign power. Stanford's lawyer said he thought they could defend any case by citing the First Amendment's protection of free speech. But the university could cover the legal costs only for professors. That's why the students' families persuaded them to keep schtum.

What was this information that US spooks considered so dangerous? Were the students proposing to read out the genetic code of smallpox, or lift the lid on some shocking conspiracy involving the president? No: they were planning to give the humdrum-sounding International Symposium on

Information Theory an update on their work on public key cryptography.

The year was 1977. If the government agency had been successful in their attempts to silence academic cryptographers, they might have prevented the internet as we know it.

To be fair, that wasn't what they had in mind. The World Wide Web was years away. And the agency's head, Admiral Bobby Ray Inman, was genuinely puzzled about the academics' motives. In his experience, cryptography – the study of sending secret messages – was of practical use only for spies and criminals. Three decades earlier, other brilliant academics had helped to win the war by breaking the Enigma code, enabling the Allies to read secret Nazi communications. Now Stanford researchers were freely disseminating information that might help adversaries in future wars to encode their messages in ways the US couldn't crack. To Inman, it seemed perverse.

His concern was reasonable. Throughout history, the development of cryptography has indeed been driven by conflict. Two thousand years ago, Julius Caesar sent encrypted messages to far-flung outposts of the Roman empire – he'd arrange in advance that recipients should simply shift the alphabet by some predetermined number. So, for example, 'jowbef Csjubjo', if you substitute all the letters with the one before them, would read 'invade Britain'.

That kind of thing wouldn't have taken the Enigma codebreakers long to crack, and encryption is typically now numerical: first, convert the letters into numbers, then perform some complicated mathematics on them. Still, the message's recipient needs to know how to unscramble the numbers by performing the same mathematics in reverse. That's known as *symmetrical* encryption. It's like securing a message with a padlock, having first given the recipient a key.

The Stanford researchers were interested in whether encryption could be *asymmetrical*. Might there be a way to send an encrypted message to someone you'd never met before, someone you didn't even know – and be confident that they, and only they, would be able to decode it?

It sounds impossible, and before 1976 most experts would have said it was. Then came the publication of a breakthrough paper by Whitfield Diffie and Martin Hellman; it was Hellman who, a year later, would defy the threat of prosecution by presenting his students' paper. That same year, three researchers at MIT – Ron Rivest, Adi Shamir and Leonard Adleman – turned the Diffie-Hellman theory into a practical technique. It's called RSA encryption, after their surnames.*

What these academics realised was that some mathematics are a lot easier to perform in one direction than another. Take a very large prime number – one that's not divisible by anything other than itself. Then take another. Multiply them together. That's simple enough, and it gives you a very, *very* large 'semi-prime' number. That's a number that's divisible only by two prime numbers.

Now challenge someone else to take that semi-prime number, and figure out which two prime numbers were multiplied together to produce it. That, it turns out, is exceptionally hard.

Public key cryptography works by exploiting this difference. In effect, an individual publishes his semi-prime number – his *public key* – for anyone to see. And the RSA algorithm allows others to encrypt messages with that number, in such a way that they can be decrypted only by

---

* As Simon Singh points out in *The Code Book* (1999), it transpired much later that British researchers working for the Government Communications Headquarters (GCHQ) had actually developed the crucial ideas of public key cryptography a few years earlier. That research was classified, and remained secret until 1997.

someone who knows the two prime numbers that produced it. It's as if you could distribute open padlocks for the use of anyone who wants to send you a message – padlocks only you can then unlock. They don't need to have your private key to protect the message and send it to you; they just need to snap shut one of your padlocks around it.

Now, in theory it's possible for someone else to pick your padlock by figuring out the right combination of prime numbers. But it takes unfeasible amounts of computing power. In the early 2000s, RSA Laboratories published some semi-primes and offered cash prizes to anyone who could figure out the primes that produced them. Someone did scoop a twenty-thousand-dollar reward – but only after getting eighty computers to work on the number non-stop for five months. Larger prizes for longer numbers went unclaimed.

No wonder Admiral Inman fretted about this knowledge reaching America's enemies. But Professor Hellman had understood something the spy chief had not. The world was changing. Electronic communication would become more important. And many private-sector transactions would be impossible if there were no way for citizens to communicate securely.

Professor Hellman was right, and you demonstrate it every time you send a confidential work email, or buy something online, or use a banking app, or visit any website that starts with 'https'. Without public key cryptography, anyone at all would be able to read your messages, see your passwords and copy your credit card details. Public key cryptography also enables websites to prove their authenticity – without it, there'd be many more phishing scams. The internet would be a very different place, and far less economically useful. Secure messages aren't just for secret agents any more: they're part of the everyday business of making your online shopping secure.

To his credit, the spy chief soon came to appreciate that the professor had a point. He didn't follow through on the threat to prosecute. Indeed, the two developed an unlikely friendship. But then, Admiral Inman was right, too – public key cryptography really does complicate his job. Encryption is just as useful to drug dealers, child pornographers and terrorists as it is to you and me when we pay for some printer ink on eBay. From a government perspective, perhaps the ideal situation would be if encryption can't be easily cracked by ordinary folk or criminals – thereby securing the internet's economic advantages – but the government can still see everything that's going on. The agency Inman headed was called the National Security Agency (NSA). In 2013, Edward Snowden released secret documents showing just how the NSA was pursuing that goal.

The debate Snowden started rumbles on. If we can't restrict encryption only to the good guys, what powers should the state have to snoop – and with what safeguards?

Meanwhile, another technology threatens to make public key cryptography altogether useless. That technology is quantum computing. By exploiting the strange ways in which matter behaves at a quantum level, quantum computers could potentially perform some kinds of calculation orders of magnitude more quickly than regular computers. One of those calculations is taking a large semi-prime number and figuring out which two prime numbers you'd have to multiply to get it. If that becomes easy, the internet becomes an open book.

Quantum computing is still in its early days. But forty years after Diffie and Hellman laid the groundwork for internet security, academic cryptographers are now racing to maintain it.

25

Double-Entry Bookkeeping

In 1495 or thereabouts, Leonardo da Vinci himself, the genius's genius, noted down a list of things to do in one of his famous notebooks. Da Vinci's To Do lists, written in mirror-writing and interspersed with sketches, are magnificent. 'Find a master of hydraulics and get him to tell you how to repair a lock, canal and mill in the Lombard manner'; 'Ask the Florentine Merchant Benedetto Portinari by what means they go on ice in Flanders'; and the deceptively brief 'Draw Milan'.

This list included the entry: 'learn multiplication from the root from Maestro Luca'. Leonardo was a big fan of Maestro Luca, better known today as Luca Pacioli. Pacioli was, appropriately enough, a Renaissance man – educated for a life in commerce, he was also a conjuror, a master of chess, a lover of puzzles, a Franciscan friar and a professor of mathematics. But today Luca Pacioli is celebrated as the most famous accountant who ever lived.

Pacioli is often called the father of double-entry bookkeeping, but he did not invent it. The double-entry system – known in the day as bookkeeping *alla Veneziana,* in the Venetian style – was being used two centuries earlier, around 1300. The

Venetians had abandoned as impractical the Roman system of writing numbers, and were instead embracing Arabic numerals. They may have also taken the idea of double-entry bookkeeping from the Islamic world, or even from India, where there are tantalising hints that such techniques date back thousands of years. Or it may have been a local Venetian invention, repurposing the new Arabic mathematics for commercial purposes.

Before the Venetian style caught on, accounts were rather basic. An early medieval merchant was little more than a travelling salesman. He had no need to keep accounts – he could simply check whether his purse was full or empty. A feudal estate needed to keep track of expenses, but the system was rudimentary: someone would be charged to take care of a particular part of the estate and would give a verbal 'account' of how things were going and what expenses had been incurred. This account would be heard by witnesses – the 'auditors', literally 'those who hear'. In English the very language of accountancy harks back to a purely oral tradition. The Chinese had written accounts, but they focused more on the problem of running a bureaucracy than running a business – in particular they weren't up to the problem of dealing with borrowing and lending.

But as the commercial enterprises of the Italian city states grew larger, more complex and more dependent on financial instruments such as loans and currency trades, the need for a more careful reckoning became painfully clear. We have a remarkable record of the business affairs of Francesco di Marco Datini, a merchant from Prato, near Florence. Datini kept accounts for nearly half a century – 1366 to 1410. They started as little more than a financial diary, but as Datini's business grew more complex, he needed something more sophisticated.

For example, late in 1394 Datini ordered wool from Mallorca, off the coast of Spain. Six months later the sheep were shorn, and several months after that, twenty-nine sacks of wool arrived in Pisa, via Barcelona. The wool was coiled into thirty-nine bales. Of these, twenty-one went to a customer in Florence, and the remaining eighteen to Datini's warehouse – arriving in 1396, over a year after the initial order. There, they were beaten, greased, combed, spun, teaseled, dyed, pressed and folded by over a hundred separate subcontractors. The final product – six long cloths – went back to Mallorca via Venice, but they didn't sell in Mallorca so were hawked in Valencia and North Africa. The last cloth was sold in 1398, nearly four years after Datini originally ordered the wool.

No wonder that Datini was so anxious and insistent on absolute clarity about inventory, assets and liabilities. He berated one befuddled associate, 'You cannot see a crow in a bowlful of milk!' and declared to another, 'You could lose your way from your nose to your mouth!' But Datini himself did not get lost in a tangle of his own financial affairs, because a decade before ordering the wool he began using the state-of-the art system of bookkeeping *alla Veneziana*.

So what, a century later, did the much-lauded Luca Pacioli add to the discipline of bookkeeping? Quite simply, in 1494, he wrote the book. And what a book it was: *Summa de Arithmetica, Geometrica, Proportioni et Proportionalita* was an enormous survey of everything that was known about mathematics in 615 large and densely typeset pages. In this colossal textbook, Pacioli included twenty-seven pages that are regarded by many as the most influential work in the history of capitalism. It was the first description of double-entry bookkeeping to be set out clearly, in detail and with plenty of examples.

Amidst the geometry and the arithmetic, it's a practical guide. Pacioli reminds his readers that they might be doing business from Antwerp to Barcelona, facing different customs and measurements in each city. 'If you cannot be a good accountant,' he warns, 'you will grope your way forward like a blind man and may meet great losses.'

Pacioli's book was sped on its way by a new technology: half a century after Gutenberg developed the movable type printing press, Venice was a centre of the printing industry. Pacioli enjoyed an impressive print run of two thousand copies, and his book was widely translated, copied and plagiarised across Europe. Double-entry bookkeeping was slow to catch on, perhaps because it was technically demanding and unnecessary for simple businesses. But after Pacioli it was always regarded as the pinnacle of the art. And as the industrial revolution unfolded, the ideas that Pacioli had set out came to be viewed as an essential part of business life; the system used across the world today is essentially the one that Pacioli described.

But what was that system? In essence Pacioli's system has two key elements. First, he describes a method for taking an inventory and then keeping on top of day-to-day transactions using two books – a rough memorandum and a tidier, more organised journal. Second, he uses a third book – the ledger – as the foundation of the system, the double-entries themselves. Every transaction is recorded twice in the ledger: for example, if you sell cloth for a ducat, you must account for both the cloth and the ducat. The double-entry system helps to catch errors – because every entry should be balanced by a counterpart. And this balance, this symmetry, seems almost divine – appealingly enough for a Renaissance mathematician.

It was during the industrial revolution that double-entry bookkeeping became seen not just as an exercise for mathematical perfectionists, but as a tool to guide practical business decisions. One of the first to see this was Josiah Wedgwood, the pottery entrepreneur. At first, Wedgwood, flush with success and fat margins, didn't bother with detailed accounts. But in 1772, Europe faced a severe recession and demand for Wedgwood's ornate crockery collapsed. His warehouses began to fill with unsold stock; his workers stood idle. How should he respond?

Faced with this crisis, Wedgwood turned to double-entry bookkeeping to understand where exactly in his business the profits were emerging, and how to expand them. He realised how much each piece of work was costing him – a deceptively simple-sounding question – and calculated that he should actually expand production and cut prices to win new customers. Others followed, and the discipline of 'management accounting' was born – an ever-growing system of metrics, benchmarks and targets that has led us inexorably to the modern world.

But in that modern world, accounting does have one more role. It's not just about making sure that basic obligations are fulfilled, as with a list of credits and debts – nor about a Venetian merchant keeping tabs on his affairs, nor even a pottery magnate trying to get on top of his costs. It's about ensuring that shareholders in a business receive a fair share of corporate profits – when only the accountants can say what those profits really are.

And here the track record is not encouraging. A string of twenty-first-century scandals – Enron, Worldcom, Parmalat and of course the financial crisis of 2007–8 – have shown us that audited accounts do not completely protect investors. A business may, through fraud or mismanagement, be on the verge of collapse. Yet we cannot guarantee that the accounts will warn us of this.

Accounting fraud is not a new game. Among the first companies to require major capital investment were the railways: they needed to raise large sums to lay tracks, long before they could hope to make a penny in profit. Unfortunately, not everyone got as rich as Cornelius Vanderbilt from these long-range investments. Britain in the 1830s and 1840s experienced 'railway mania'. Many speculators ploughed their savings into proposed new routes that never delivered the promised financial returns – or in some cases were never built at all. When the railway company in question could not pay the expected dividends, they kept the bubble inflated by simply faking their accounts. As a physical investment the railways were a triumph, but as a financial punt they were often a disaster. The bubble in railway stocks and bonds had collapsed in ignominy by 1850.

Perhaps the railway investors should have read up on their Geoffrey Chaucer, writing around the same time as Francesco Datini, the merchant of Prato. In Chaucer's Shipman's Tale, a rich merchant is too tied up with his accounts to notice that his wife is being wooed by a clergyman. Nor do those accounts rescue him from an audacious con: the clergyman borrows the merchant's money, gives it to the merchant's wife – thus buying his way into her bed with her own husband's cash – and then tells the merchant he's repaid the debt and to ask his wife where the money is.

Accountancy is a powerful financial technology – but it does not protect us from outright fraud, and it may well lure us into complacency. As the neglected wife tells her rich husband, his nose buried in his accounts: 'the devil take all such reckonings!'

26

Limited Liability Companies

Nicholas Murray Butler was one of the thinkers of his age: a philosopher, Nobel Peace Prize winner, president of Columbia University. In 1911, someone asked Butler to name the most important invention of the industrial era. Steam, perhaps? Electricity? No, he said: they would both 'be reduced to comparative impotence' without something else – something he called 'the greatest single discovery of modern times'. That something? The limited liability corporation.

It seems odd to say the corporation was 'discovered'. But it didn't just appear from nowhere. The word 'incorporate' means take on bodily form – not a physical body, but a legal one. In the law's eyes, a corporation is something different from the people who own it, or run it, or work for it. And that's a concept lawmakers had to dream up. Without laws saying that a corporation can do certain things – like own assets, or enter into contracts – the word would be meaningless.

There were precursors to modern corporations in ancient Rome, but the direct ancestor of today's corporations was born in England, on New Year's Eve, in 1600. Back then, creating a corporation didn't simply involve filing some routine forms – you needed a royal charter. And you couldn't incorporate with the general aim of doing business and making profits. A corporation's charter specifically said what it was allowed to do, and often also stipulated that nobody else was allowed to do it.

The legal body created that New Year's Eve was charged with handling all of England's shipping trade east of the Cape of Good Hope. Its shareholders were 218 merchants. Crucially, and unusually, the charter granted those merchants *limited liability* for the company's actions.

Why was that so important? Because otherwise, investors were personally liable for everything the business did. If you partnered in a business which ran up debts it couldn't pay, its debtors could come after you – not just for the value of your investment, but for everything you owned.

That's worth thinking about: whose business might you be willing to invest in, if you knew that it could lose you your home, and even land you in prison? Perhaps a close family member's; at a push, a trusted friend's. Someone you knew well enough, and saw often enough, to notice if they were behaving suspiciously. The way we invest today – buying shares in companies whose managers we will never meet – would be unthinkable. And that would severely limit the amount of capital a business venture could raise.

Back in the 1500s, perhaps that wasn't much of a problem. Most business was local, and personal. But handling England's trade with half the world was a weighty undertaking. The corporation Queen Elizabeth created was called the East India Company. Over the next two centuries, it grew to look less like a trading business than a colonial government. At its peak, it ruled 90 million Indians. It employed an army of 200,000 soldiers. It had a meritocratic civil service. It issued its own coins.

And the idea of limited liability caught on. In 1811, New York state introduced it – not as a royal privilege, but for any manufacturing company. Other states and countries followed, including the world's leading economy, Britain, in 1854. Not everyone approved: *The Economist* magazine was sniffy, pointing out that if people wanted limited liability they could agree it through private contracts.

As we've seen, nineteenth-century industrial technologies such as railways and electricity grids needed capital – lots of capital. And that meant either massive government projects – not fashionable then – or limited liability companies.

And limited liability companies proved their worth. Soon *The Economist* was gushing that the unknown inventors of limited liability deserved 'a place of honour with Watt, Stephenson and other pioneers of the industrial revolution'.

But as the railway mania demonstrated, the limited liability corporation has its problems. One of them was obvious to the father of modern economic thought, Adam Smith. In *The Wealth of Nations*, in 1776, Smith dismissed the idea that professional managers would do a good job of looking after shareholders' money: 'it cannot well be expected that they should watch over it with the same anxious vigilance with which partners in a private copartnery frequently watch over their own', he wrote.

In principle, Smith was right. There's always a temptation for managers to play fast and loose with investors' money. We've evolved corporate governance laws to try to protect shareholders, but as we've seen they haven't always succeeded.

And corporate governance laws generate their own tensions. Consider the fashionable idea of 'corporate social responsibility' – under which a company might donate to charity, or decide to embrace labour or environmental standards above what the law demands. In some cases, that's smart brand-building, and it pays off in higher sales. In others, perhaps, managers are using shareholders' money to buy social status or a quiet life. For that reason, the economist Milton Friedman argued that 'the social responsibility of business is to maximise its profits'. If it's legal, and it makes money, they should do it. And if people don't like it, don't blame the company – change the law.

The trouble is that companies can influence the law, too. They can fund lobbyists. They can donate to electoral candidates' campaigns. The East India Company quickly learned the value of maintaining cosy relationships with British politicians, who duly bailed it out whenever it got into trouble. In 1770, for example, a famine in Bengal clobbered the company's revenue. British legislators saved it from bankruptcy, by exempting it from tariffs on tea exports to the American colonies. Which was, perhaps, shortsighted on their part: it eventually led to the Boston Tea Party, and the American Declaration of Independence. You could say the United States owes its existence to excessive corporate influence on politicians.

And arguably, corporate power is even greater today, for a simple reason: in a global economy, corporations can threaten to move offshore. The shipping container and the barcode have underpinned global supply chains, giving companies the ability to locate key functions wherever they wish. When Britain's lawmakers eventually grew tired of the East India Company's demands, they had the ultimate sanction – in 1874, they revoked its charter. A government dealing with a modern multinational must exercise influence far more carefully than that.

We often think of ourselves as living in a world where free-market capitalism is the dominant force. Few want a return to the command economies of Mao or Stalin, where hierarchies, not markets, decided what to produce. And yet hierarchies, not markets, are exactly how decisions are made *within* companies: when a receptionist or an accounts payable clerk makes a decision, they're not doing so because the price of soy beans has risen. They're following orders from the boss. In the United States, bastion of free-market capitalism, about half of all private-sector employees work for companies with a payroll of at least five hundred.

Some argue that companies have grown too big, too influential. In 2016, Pew Research asked Americans if they thought the economic system is 'generally fair', or 'unfairly favours powerful interests' – by two-to-one, unfair won. Even *The Economist* worries that regulators are now too timid about exposing market-dominating companies to a blast of healthy competition.

So there's plenty to worry about. But while we worry, let's also remember what the limited liability company has done for us. By helping investors pool their capital without taking unacceptable risks, it enabled big industrial projects, stock markets and index funds. It played a foundational role in creating the modern economy.

27

Management Consulting

The place: a textile plant near Mumbai, India. The time: 2008. The scene? Chaos. Rubbish is piled up outside the building – and almost as much inside, for that matter. There are piles of flammable junk, and uncovered containers of chemicals. The yarn is scarcely neater: it is at least bundled up and protected in white plastic bags, but the inventory is scattered around the plant in unmarked piles.

Such shambolic conditions are typical in the Indian textile industry, and that presents an opportunity. A team of researchers from Stanford University and the World Bank is about to conduct a novel experiment: they're going to send in a team of management consultants to tidy up some of these companies, but not others. Then they'll track what happens to their profits. This will be a rigorous, randomised controlled trial. It will conclusively tell us whether the management consultants are worth their fees.

That question has often been raised over the years, with a sceptical tone. If managers tend to have a bad reputation, what should we make of the people who tell managers how to manage? Picture a management consultant: what comes to mind? Perhaps a young, sharp-suited graduate, earnestly gesturing to a bulleted PowerPoint presentation that reads something like: 'holistically envisioneer client-centric deliverables'.

Okay, I got that from an online random buzzword generator. But you get the idea. The industry battles a stereotype of charging exorbitant fees for advice that, on close inspection, turns out to be either meaningless or common sense. Managers who bring in consultants are often accused of being blinded by jargon, implicitly admitting their own incompetence, or seeking someone else to blame for unpopular decisions.

Still, it's big business. The year after Stanford and the World Bank started their Indian study, the UK government alone spent £1.8 billion on management consultants. Globally, consulting firms charge their clients a total of about 125 billion dollars.

Where did this strange industry begin?

There's a noble way to frame its origins: economic change creates a new challenge, and visionary men of business provide a solution. In the late nineteenth century, the US economy was expanding fast, and thanks to the railway and telegraph it was also integrating, becoming more of a national market and less of a collection of local ones. Company owners began to realise that there were huge rewards to be had for companies that could bestride this new national stage. So began an unprecedented wave of mergers and consolidations: companies swallowed each other up, creating giant household names: US Steel, General Electric, Heinz, AT&T. Some employed over 100,000 people. And that was the challenge: nobody had ever tried to manage such vast organisations before.

In the late 1700s Josiah Wedgwood had shown that

double-entry bookkeeping techniques could help business owners understand where they were making money, and what steps they might take to make more. But using accounts to actually manage a large corporation would require a new approach.

Enter a young professor of accountancy by the name of James McKinsey. McKinsey's breakthrough was a book published in 1922, with the not-entirely-thrilling title *Budgetary Control*. But for corporate America, *Budgetary Control* was revolutionary. Rather than using traditional historical accounts to provide a picture of how a business had been doing over the past year, McKinsey proposed drawing up accounts for an imaginary corporate future. These future accounts would set out a business's plans and goals, broken down department by department. And later, when the actual accounts were drawn up they could be compared to the plan, which could then be revised. McKinsey's method helped managers take control, setting out a vision for the future rather than simply reviewing the past.

McKinsey was a big character: tall and fond of chomping cigars, ignoring his doctor's advice. His ideas caught on with remarkable speed: by the mid-1930s, he was hiring himself out at five hundred dollars a day – about twenty-five thousand dollars in today's money. And, since he was busy, he took on employees; if he didn't like a report they wrote, he'd hurl it in the bin. 'I have to be diplomatic with our clients,' he told them. 'But I don't have to be diplomatic with you bastards!'

And then, at the age of forty-eight, James McKinsey died of pneumonia. But under his lieutenant, Marvin Bower, McKinsey & Company thrived. Bower was a particular man. He insisted that the men who worked for him wore a dark suit, a starched white shirt, and, until the 1960s, a hat. McKinsey & Co, he said, was not a business but a 'practice';

it didn't take on jobs, it took on 'engagements'; it was not a company, it was a 'firm'. Eventually it simply became known as 'The Firm'. Duff McDonald wrote a history of The Firm, arguing that its advocacy of scientific approaches to management transformed the business world. It acquired a reputation as perhaps the world's most elite employer. *The New Yorker* once described McKinsey's young Ivy League hires parachuting into companies around the world as 'a SWAT team of business philosopher-kings'.

But hold on: why don't company owners simply employ managers who've studied those scientific approaches themselves? There aren't many situations where you'd hire someone to do a job, and also hire expensive consultants to advise them how to do it. What accounts for why companies like McKinsey gained such a foothold in the economy?

Part of the explanation is surprising: government regulators cleared a niche for them. The Glass-Steagall Act of 1933 was a far-reaching piece of American financial legislation. Among many provisions, Glass-Steagall made it compulsory for investment banks to commission independent financial research into the deals they were brokering; fearing conflicts of interest, Glass-Steagall forbade law firms, accountancy firms and the banks themselves from conducting this work. In effect, the Glass-Steagall Act made it a legal requirement for banks to hire management consultants. For a follow-up, in 1956 the Justice Department banned the emerging computer giant IBM from providing advice about how to install or use computers: another business opportunity for the management consultants.

Minimising conflicts of interest was a noble aim, but it hasn't worked out well. A few years after leaving the firm, McKinsey's long-serving boss, Rajat Gupta, was convicted and imprisoned for insider trading. McKinsey also employed

Enron's Jeff Skilling, and then was paid well for advising him, before quietly fading into the background while Enron collapsed and Skilling went to jail.

Here's another argument for employing management consultants: ideas on management evolve all the time, so maybe it's worth getting outsiders in periodically for a burst of fresh thinking? Clearly that can work. But often it doesn't. Instead, the consultants continually find new problems to justify their continued employment – like leeches, attaching themselves and never letting go. It's a strategy known as 'land and expand'. One UK government ministry recently admitted that 80 per cent of its supposedly temporary consultants had been working there for more than a year – some for up to nine years. Needless to say, it would have been much cheaper to employ them as civil servants.

No doubt the consultancy firms will claim that their expertise is giving the taxpayer value for money. Which brings us back to India, and that randomised controlled trial. The World Bank hired the global consulting firm Accenture to put some structure into these jumbled Mumbai textile factories, instituting new routines: preventative maintenance, proper records, systematic storing of spares and inventory, and the recording of quality defects. And did it work?

It did. Productivity jumped by 17 per cent – easily enough to pay Accenture's consulting fees. We shouldn't conclude from this study that cynicism about management consulting is always misplaced. These factories were, after all, what a jargon-filled PowerPoint presentation might call 'low-hanging fruit'. But it's scientific proof of one thing, at least: as so often in life, when an idea is used simply, and humbly, it can pay dividends.

28

Intellectual Property

In January 1842, Charles Dickens arrived on American shores for the first time. He was greeted like a rock star in Boston, Massachusetts, but the great novelist was a man with a cause: he wanted to put an end to cheap, sloppy pirated copies of his work in the US. They circulated with impunity because the United States granted no copyright protection to non-citizens. In a bitter letter to a friend, Dickens compared the situation to being mugged and then paraded through the streets in ridiculous clothes. 'Is it tolerable that besides being robbed and rifled, an author should be forced to appear in any form – in any vulgar dress – in any atrocious company ... ?'

It was a powerful and melodramatic metaphor; from Dickens, what else would one expect? But the truth is that the case for what Dickens was demanding – legal protection for ideas that otherwise could be freely copied and adapted – has never been quite so clear cut.

Patents and copyright grant a monopoly, and monopolies are bad news. Dickens's British publishers will have charged as much as they could get away with for copies of *Bleak House*, and cash-strapped literature lovers simply had to go without.

But these potential fat profits encourage new ideas. It took Dickens a long time to write *Bleak House*. If other British publishers could have ripped it off like the Americans, perhaps he wouldn't have bothered.

So intellectual property reflects an economic trade-off – a balancing act. If it's too generous to the creators then good ideas will take too long to copy, adapt and spread. If it's too stingy then maybe we won't see the good ideas at all.

One might hope that the trade-off would be carefully weighed up by benevolent technocrats, but it has always been coloured by politics. The British legal system strongly protected the rights of British authors and British inventors in the 1800s because the UK was then – as it remains – a powerful force in world culture and innovation. But in Dickens's day, American literature and American innovation were in their infancy. The US economy was in full-blown copying mode: they wanted the cheapest possible access to the best ideas that Europe could offer. American newspapers filled their pages with brazen copying – alongside their attacks on the interfering Mr Dickens.

A few decades later, when American authors and inventors spoke with a more powerful voice, America's lawmakers began to take an increasingly fond view of the idea of intellectual property. Newspapers, once opposed to copyright, began to rely upon it. The United States finally began to respect international copyright in 1891, half a century after Dickens's campaign. And we can expect to see a similar transition in developing countries today: the less they copy other ideas and the more they create of their own, the more they protect ideas themselves. There's been a lot of movement in a brief time: China didn't have a system of copyright at all until 1991.

The modern form of intellectual property originated, like

so many things, in fifteenth-century Venice. Venetian patents were explicitly designed to encourage innovation. They applied consistent rules: the inventor would automatically receive a patent if the invention was useful; the patent was temporary, but while it lasted it could be sold, transferred or even inherited; the patent would be forfeited if it wasn't used; and the patent would be invalidated if the invention proved to be closely based on some previous idea. These are all very modern ideas.

And they soon created very modern problems. During the British industrial revolution, for example, the great engineer James Watt figured out a better way to design a steam engine. He spent months developing a prototype, but then put even more effort into securing a patent. His influential business partner, Matthew Boulton, even got the patent extended by lobbying Parliament. Boulton and Watt used it to extract licensing fees and crush rivals – for example, Jonathan Hornblower, who made a superior steam engine yet found himself ruined and imprisoned.

The details may have been grubby, but surely Watt's famous invention was worth it? Maybe not. The economists Michele Boldrin and David Levine argue that what truly unleashed steam-powered industry was the *expiry* of the patent, in 1800, as rival inventors revealed the ideas they had been sitting on for years. And what happened to Boulton and Watt, once they could no longer sue those rivals? They flourished anyway. They redirected their attention from litigation towards the challenge of producing the best steam engines in the world. They kept their prices as high as ever, and their order books swelled.

Far from incentivising improvements in the steam engine, then, the patent actually delayed them. Yet since the days of Boulton and Watt, intellectual property protection has grown *more* expansive, not less so. Copyright terms are growing ever longer: in the US, they were originally fourteen years, renewable once. They now last seventy years after the death of the author – typically more than a century. Patents have become broader – they're being granted on vague ideas – for example, Amazon's 'one-click' US patent protects the not-entirely-radical idea of buying a product on the internet by clicking only one button. The US intellectual property system now has a global reach, thanks to the inclusion of intellectual property rules in what tend to be described as 'trade agreements'. And more and more things fall under the scope of intellectual property – for example plants, buildings, software and even the look and feel of a restaurant chain have all been brought into its domain.

These expansions are hard to justify, but easy to explain: intellectual property is very valuable to its owners, which justifies the cost of employing expensive lawyers and lobbyists. Meanwhile, the costs of the restrictions are spread widely over people who barely notice it. The likes of Matthew Boulton and Charles Dickens have a strong incentive to lobby aggressively for more draconian intellectual property laws – while the disparate buyers of steam engines and *Bleak House* are unlikely to manage to organise a strong political campaign to object.

The economists Boldrin and Levine have a radical response to this problem: scrap intellectual property altogether. There are, after all, other rewards for inventing things – getting a 'first mover' advantage over your competitors, establishing a strong brand, or enjoying a deeper understanding of what makes a product work. In 2014, the electric car company Tesla opened up access to its patent archive in an effort to expand the industry as a whole, calculating that Tesla would benefit from that.

For most economists, scrapping intellectual property entirely is going too far. They point to important cases – for example, new medicines – where the costs of invention are enormous and the costs of copying are trivial. But those who defend intellectual property protections still tend to argue that right now they're too broad, too long and too difficult to challenge. Narrower, briefer protection for authors and inventors would restore the balance, and still give plenty of incentive to create new ideas.

Charles Dickens himself eventually discovered that there's a financial upside to weak copyright protection. A quarter of a century after his initial visit to the United States, Dickens returned. His family was ruinously expensive and he needed to make some money. And he reckoned that so many people had read cheap knock-offs of his stories that he could cash in on his fame with a lecture tour. He was absolutely right: off the back of pirated copies of his work, Charles Dickens made a fortune as a public speaker, many millions of dollars in today's terms. Perhaps the intellectual property was worth more when given away.

29

The Compiler

One, zero, zero, zero, one, zero, one, one. Zero, one, one . . .

That's the language of computers. Every clever thing your computer does – make a call, search a database, play a game – comes down to ones and zeroes. Actually, that's not quite true. It comes down to the presence or absence of a current in tiny transistors on a semiconductor chip. The zero or one merely denotes if the current is off or on.

Thankfully, we don't have to program computers in zeroes and ones. Imagine how difficult that would be. Microsoft Windows, for example, takes up 20 gigabytes of space on my hard drive. That's 170 billion ones and zeroes. Print them out and the stack of A4 paper would be 4 kilometres high. Now imagine you had to work through those pages, setting every transistor manually. We'll ignore how fiddly this would be: transistors measure just billionths of a metre. If it took a second to flip each switch, installing Windows would take five thousand years.

Early computers really did have to be programmed rather like this. Consider the Automatic Sequence Controlled

Calculator, later known as the Harvard Mark 1. It was a 15-metre-long, 2.5-metre-high concatenation of wheels and shafts and gears and switches. It contained 530 miles of wires. It whirred away under instruction from a roll of perforated paper tape, like a player piano. If you wanted it to solve a new equation, you had to work out which switches should be on or off, which wires should be plugged in where. Then you had to flip all the switches, plug in all the wires and punch all the holes in the paper tape. Programming it was a challenge to stretch the mind of a mathematical genius; it was also tedious, repetitive, error-prone manual labour.

Four decades on from the Harvard Mark 1, more compact and user-friendly machines like the Commodore 64 were finding their way into schools. If you're around my age, you may remember the childhood thrill of typing this:

```
10 print 'hello world';
20 goto 10
```

And, lo – 'hello world' would fill the screen, in chunky, low-resolution text. You'd instructed the computer in words that were recognisably, intuitively human – and the computer had understood. It seemed like a minor miracle. If you ask why computers have progressed so much since the Mark 1, one reason is certainly the ever-tinier components. But it's also unthinkable that computers could do what they do if programmers couldn't write software like Windows in human-like language, and have it translated into the ones and zeroes, the currents or not-currents, that ultimately do the work.

The thing that began to make that possible was called a compiler. And the story of the compiler starts with a woman called Grace Hopper.

Nowadays there's much discussion about how to get more women into careers in tech. In 1906, when Grace was born, not many people cared about gender equality in the jobs market. Fortunately for Grace, among those who did was her father, a life insurance executive: he didn't see why his daughters should get less of an education than his son. Grace went to a good school, and turned out to be brilliant at maths. Her grandfather was a Rear Admiral, and her childhood dream was to join the Navy, but girls weren't allowed. She settled for becoming a professor instead.

Then, in 1941, the attack on Pearl Harbor dragged America into the Second World War. Male talent was called away. The Navy started taking women. Grace signed up at once.

If you're wondering what use the Navy had for mathematicians, consider aiming a missile. At what angle and direction should you fire? The answer depends on many things: how far away the target is, the temperature and humidity, and the speed and direction of the wind. The calculations involved aren't complex, but they were time-consuming for a human 'computer' – someone with a pen and paper. Perhaps there was a faster way. As Lieutenant (junior grade) Hopper graduated Midshipmen's School in 1944, the Navy was intrigued by the potential of an unwieldy contraption recently devised by Howard Aiken, a professor at Harvard. It was the Mark 1. The Navy sent Hopper to help Aiken work out what it could do.

Aiken wasn't thrilled to have a female join the team, but soon Hopper impressed him enough that he asked her to write the operating manual. Figuring out what it should say involved plenty of trial and error. More often than not, the Mark 1 would grind to a halt soon after starting – and there was no user-friendly error message. Once it was because a moth had flown into the machine, which gave us the modern term 'debugging'. More likely, the bug was metaphorical – a switch flipped wrongly, a mispunched hole in the paper tape. The detective work was laborious and dull.

Hopper and her colleagues started filling notebooks with bits of tried-and-tested, reuseable code. By 1951, computers had advanced enough to store these chunks – they were called 'subroutines' – in their own memory systems. Hopper was then working for a company called Remington Rand. She tried to convince her employer to let programmers call up these subroutines in familiar words – to say things like 'subtract income tax from pay', instead of, as Hopper put it, 'trying to write that in octal code or using all kinds of symbols'.

Hopper later claimed that 'No one thought of that earlier because they weren't as lazy as I was.' That's tongue-in-cheek self-deprecation – she was famed for hard work. But it does have a kernel of truth: the idea Hopper called a 'compiler' involved a trade-off. It made programming quicker, but the resulting programs ran more slowly. And that's why Remington Rand wasn't interested. Every customer had their own, bespoke requirements for their shiny new computing machine. It made sense, Remington Rand thought, for the company's experts to program them as efficiently as they could.

Hopper wasn't discouraged: she simply wrote the first compiler in her spare time. And others loved how it helped them to think more clearly. One impressed customer was an engineer called Carl Hammer, who used it to attack an equation his colleagues had struggled with for months – he wrote twenty lines of code and solved it in a day. Like-minded programmers all over the US started sending Hopper new chunks of code; she added them to the library for the next release. In effect, she was single-handedly pioneering open-source software.

Hopper's compiler evolved into one of the first programming languages, COBOL; more fundamentally, it paved the way for the now-familiar distinction between hardware and software. With one-of-a-kinds like the Harvard Mark 1, software *was* hardware: no pattern of switches would also work on some other machine, which would be wired completely differently. But if a computer can run a compiler, it can also run any program that uses it.

More and more layers of abstraction have since come to separate human programmers from the nitty-gritty of physical chips. And each one has taken a further step in the direction Grace Hopper realised made sense: freeing up programmer brainpower to think about concepts and algorithms, not switches and wires.

Hopper had her own views of why colleagues resisted it, at first – and it wasn't because they cared about making programs run more quickly. No, they enjoyed the prestige of being the only ones who could communicate with the godlike computer on behalf of the mere mortal who'd just purchased it. The 'high priests', Hopper called them.

Hopper thought anyone should be able to program. Now, anyone can. And computers are far more useful because of it.

V

WHERE DO INVENTIONS
COME FROM?

Many books have tried to solve the puzzle of how innovation happens. The sheer range of answers is telling. Joel Mokyr's *A Culture of Growth* looks at the huge forces in the background. Mokyr emphasises the political fragmentation of Enlightenment Europe, which left intellectuals free to move around, escaping persecution and seeking patrons. Steven Johnson's *Where Good Ideas Come From* zooms in closer, looking at the networks of people who share ideas from the coffee houses of the 1650s to Silicon Valley today. Keith Sawyer's *Explaining Creativity* looks more closely still, drawing on ideas in neuroscience and cognitive psychology. There are countless other perspectives on the question.

This book doesn't focus on the question of how inventions come into being – it's more interested in what inventions do to the social and economic structures that surround us. But almost in passing, we've already learned a lot about where inventions come from.

There's the demand-driven invention: we don't know who invented the plough, but we do know that it was a response to a changing world – foraging nomads didn't suddenly invent the technology and then take up agriculture in order to use it. Another example is barbed wire: everybody could see the need for it. Joseph Glidden produced the most practical version of many competing designs, but we know little about the details of his creative process. It seems to have been rather mundane: after all, in retrospect the design was obvious enough. Glidden just realised it first.

On the flip side, there's the supply-push invention. Betty Cronin worked for a company, Swanson, which had made good money supplying preserved rations to American troops in the Second World War. Now it had the capacity and the technology, but it needed to find a new market: the frozen TV dinner was the result of this hunt for profits.

There's invention by analogy: Sergey Brin and Larry Page developed their search algorithm drawing inspiration from academic citations; Joseph Woodland developed the barcode as he dragged his fingers through sand and pondered Morse code.

That said, the barcode itself was invented independently several times. The sticking point was the internal politics of the US retailing industry. This reminds us that there's more to an invention than the inventing of it. It wouldn't be entirely wrong to say that Malcom McLean 'invented' the shipping container, but it's more illuminating to describe the obstacles he had to overcome in getting the system to take off.

The truth is that even for a single invention, it's often hard to pin down a single person who was responsible – and it's even harder to find a 'eureka' moment when the idea all came together. Many of the inventions in this book have many parents – they often evolved over decades or centuries. The honest answer to the question, 'where do inventions come from?' is 'almost anywhere you can imagine'.

30

The iPhone

On 9 January 2007, the most iconic entrepreneur on the planet announced something new – a product that was to become the most profitable in history.

It was the iPhone. There are many ways in which the iPhone has defined the modern economy. There is the sheer profitability of the thing, of course: indeed, only two or three companies in the world make as much money as Apple does on the iPhone alone. There is the fact that it created a new product category: the smartphone. The iPhone and its imitators represent a product that did not exist ten years ago but that is now an object of desire for most of humanity. There's the way the iPhone transformed other markets – for software, for music and for advertising.

But those are just the obvious facts about the iPhone. And when you delve deeper, the tale is a surprising one. We give credit to Steve Jobs and other leading figures in Apple – his early partner Steve Wozniak, his successor Tim Cook, his visionary designer Jony Ive – but some of the most important actors in this story have been forgotten.

Ask yourself: what actually makes an iPhone an iPhone?

It's partly the cool design, the user interface, the attention to detail in the way the software works and the hardware feels. But underneath the charming surface of the iPhone are some critical elements that made it, and all the other smartphones, possible.

The economist Mariana Mazzucato has made a list of twelve key technologies that make smartphones work. One: tiny microprocessors. Two: memory chips. Three: solid state hard drives. Four: liquid crystal displays. Five: lithium-based batteries. That's the hardware.

Then there are the networks and the software.

So, continuing to count: Six: fast-Fourier-transform algorithms. These are clever bits of maths that make it possible to swiftly turn analogue signals such as sound, visible light and radio waves into digital signals that a computer can handle.

Seven – and you might have heard of this one – the internet. A smartphone isn't a smartphone without the internet.

Eight: HTTP and HTML, the languages and protocols that turned the hard-to-use internet into the easy-to-access World Wide Web. Nine: cellular networks. Otherwise your smartphone not only isn't smart, it's not even a phone. Ten: Global Positioning Systems or GPS. Eleven: The touch screen. Twelve: Siri, the voice-activated artificial intelligence agent.

All of these technologies are important components of what makes an iPhone, or any smartphone, actually work. Some of them are not just important, but indispensable. But when Mariana Mazzucato assembled this list of technologies, and reviewed their history, she found something striking. The foundational figure in the development of the iPhone wasn't Steve Jobs. It was Uncle Sam. Every single one of these twelve vital technologies was supported in significant ways by governments – often the American government.

A few of these are famous. Many people know, for

example, that the World Wide Web owes its existence to the work of Tim Berners-Lee. Berners-Lee was a software engineer employed at CERN, the particle physics research centre in Geneva that is funded by governments across Europe. And the internet itself started as ARPANET – an unprecedented network of computers funded by the US Department of Defense in the early 1960s. GPS, of course, was a pure military technology, developed during the Cold War and opened up to civilian use only in the 1980s.

Other examples are less famous, though scarcely less important.

The fast-Fourier-transform is a family of algorithms that have made it possible to move from a world where the telephone, the television and the gramophone worked on analogue signals, to a world where everything is digitised and can therefore be dealt with by computers such as the iPhone. The most common such algorithm was developed from a flash of insight by the great American mathematician John Tukey. What was Tukey working on at the time? You've guessed it: a military application. Specifically, he was on President Kennedy's Scientific Advisory Committee in 1963, trying to figure out how to detect when the Soviet Union was testing nuclear weapons.

Smartphones wouldn't be smartphones without their touch screens – but the inventor of the touch screen was an engineer named E.A. Johnson, whose initial research was carried out while Johnson was employed by the Royal Radar Establishment, a stuffily named agency of the British government. The work was further developed at CERN – those guys again. Eventually multi-touch technology was commercialised by researchers at the University of Delaware in the United States – Wayne Westerman and John Elias, who sold their company to Apple itself. Yet even at that late stage in the game, governments played their part: Wayne Westerman's research fellowship was funded by the US National Science Foundation and the CIA.

Then there's the girl with the silicon voice, Siri.

Back in the year 2000, seven years before the first iPhone, the US Defense Advanced Research Projects Agency, DARPA, commissioned the Stanford Research Institute to develop a kind of proto-Siri, a virtual office assistant that might help military personnel to do their jobs. Twenty universities were brought into the project, furiously working on all the different technologies necessary to make a voice-activated virtual assistant a reality. In 2007, the research was commercialised as a start-up, Siri Incorporated – and it was only in 2010 that Apple stepped in to acquire the results for an undisclosed sum.

As for hard drives, lithium-ion batteries, liquid crystal displays and semiconductors themselves – there is a similar story to be told. In each case there was scientific brilliance and plenty of private-sector entrepreneurship. But there were also wads of cash thrown at the problem by government agencies – usually US government agencies, and for that matter usually some arm of the US military. Silicon Valley itself owes a great debt to Fairchild Semiconductor – the company that developed the first commercially practical integrated circuits. And Fairchild Semiconductor, in its early days, depended on military procurement.

Of course, the US military didn't make the iPhone. CERN did not create Facebook or Google. These technologies that so many people rely on today were honed and commercialised by the private sector. But it was government funding and government risk-taking that made all these things possible. That's a thought to hold on to as we ponder the technological challenges ahead in fields such as energy and biotechnology.

Steve Jobs was a genius, there's no denying that. One of his remarkable side projects was the animation studio Pixar, which changed the world of movies when it released the digitally animated film, *Toy Story*.

Even without the touch screen and the internet and the fast-Fourier-transform, Steve Jobs might well have created something wonderful. But it would not have been a world-shaking technology like the iPhone. More likely it would, like Woody and Buzz, have been an utterly charming toy.

31

Diesel Engines

It was 10 p.m. Rudolf Diesel had finished dinner and retired to his cabin aboard the S.S. *Dresden*, travelling from Belgium across the English Channel. His night clothes were laid out on his bed, but Diesel did not change into them. The inventor of the engine that bears his name was thinking about his heavy debts, and the interest payments that would soon come due. He couldn't afford them. In his diary, today's date – 29 September 1913 – was marked with an ominous 'X'.

Before the trip, Diesel had gathered what cash he could and stuffed it into a bag, together with documents laying bare the financial mess he was in. He gave the bag to his wife, telling her not to open it until a week had passed; she seems not to have suspected anything. Diesel stepped outside his cabin. He removed his coat, folded it and laid it neatly on the ship's deck. He looked over the railings, at the black and swirling waters below. And he jumped.

Or did he? While this seems the most plausible account of Rudolf Diesel's final moments, it remains an assumption. Conspiracy theorists have speculated that Diesel was assisted overboard. But who might have had an interest in the impecunious inventor's demise? Two possible candidates have been fingered. The conspiracies may well be entirely baseless; nonetheless, they help us to understand the economic significance of the engine Diesel invented in 1892.

For context, rewind another twenty years, to 1872 and industrial economies in which steam supplied the power for trains and factories, but urban transport depended on horses. That autumn, equine flu brought US cities to a standstill. Grocery store shelves were bare; saloons ran out of beer; garbage piled up in the streets. A city of half a million people might have 100,000 horses, and each one liberally coated the streets with 35 pounds of manure and a gallon of urine every day. An affordable, reliable, small-scale engine that could replace the horse would be a godsend.

The steam engine was one candidate: steam-powered cars were coming along nicely. Another was the internal combustion engine, early versions of which ran on petrol, gas, or even gunpowder. But when Rudolf Diesel was a student, both types of engine were woefully inefficient: they converted only around 10 per cent of heat into useful work.

The young Diesel's life was changed by a lecture on thermodynamics at the Royal Bavarian Polytechnic of Munich, which discussed the theoretical limits to the efficiency of an engine. The 10 per cent efficiency achieved in practice looked very low by the standards of the lecturer's theorems, and Diesel became obsessed with the goal of making an engine that came as close as possible to converting all heat into work. Of course, in practice perfect efficiency is impossible – but his first working engine was over 25 per cent efficient, which was more than twice as good as the state of the art at the time. Today, the best diesel engines top 50 per cent.

Petrol engines work by compressing a mixture of fuel and air, then igniting it using a spark plug. But compress the mixture too much and it can prematurely self-ignite, which causes destabilising engine knock. Diesel's invention avoids that problem by compressing only the air, and more so, making it hot enough to ignite the fuel when it's injected. This allows the engine to be more efficient: the higher the compression, the less fuel is needed. Anyone who's researched buying a car will be familiar with the basic trade-off of a diesel engine – they tend to be more expensive to buy, but more economical to run.

Unfortunately for Rudolf, in early versions of his engine these efficiency gains were outweighed by reliability issues. He faced a steady stream of refund demands from unhappy customers; it was this that dug the inventor into the financial hole from which he never managed to escape. It's ironic: the inventor of one of the most straightforwardly practical machines in the modern economy was motivated by an inspiring lecture rather than by money, which is just as well, since he failed to make any.

Still, he kept working at his engine, and it kept getting better. Other advantages became apparent. Diesel engines can use a heavier fuel than petrol engines – specifically, a heavier fuel that's become known as 'diesel'. As well as being cheaper than petrol to refine from crude oil, diesel also gives off fewer fumes, so it's less likely to cause explosions. This made it particularly attractive for military transport: after all, you don't want your bombs going off accidentally. By 1904, Diesel had got his engines into France's submarines.

This brings us to the first conspiracy theory around Rudolf Diesel's death. In the Europe of 1913, the drumbeats of impending war were quickening and the cash-strapped German was en route to London. One newspaper headline luridly speculated: 'Inventor Thrown Into the Sea to Stop Sale of Patents to British Government'.

It was only after the First World War that Diesel's invention really began to realise its commercial potential, in heavier-duty transport applications than cars. The first diesel-powered trucks appear in the 1920s and trains in the 1930s; by 1939 a quarter of global sea trade was fuelled by diesel. After the Second World War, ever more powerful and efficient diesel engines led to ever more enormous ships. Diesel's invention, quite literally, is the engine of global trade.

Fuel accounts for about 70 per cent of the costs of shipping goods around the world. You can see why the scientist Vaclav Smil reckons that if globalisation had been powered by steam, rather than diesel, trade would have grown much more slowly than it did.

The economist Brian Arthur isn't so sure about that. Arthur views the rise of the internal combustion engine over the last century as an example of 'path dependence' – a self-reinforcing cycle in which existing investments and infrastructure mean we keep doing things in a certain way, even if we'd do them differently if only we could start from scratch. As late as 1914, Arthur argues, steam was at least as viable as crude oil for powering cars – but the growing influence of the oil industry ensured that much more money was going into improving the internal combustion engine than the steam engine. With equal investment in research and development, who knows where breakthroughs might have happened; perhaps today we'd be driving next-generation steam-powered cars.

Alternatively, if Rudolf Diesel had had his way, perhaps the global economy would run on peanuts.

Diesel's name has become synonymous with a crude oil derivative, but he designed his engine to use a variety of fuels, from coal dust to vegetable oils. In 1900, at the Paris World Fair, he demonstrated a model based on peanut oil and, as the years went by, he became something of an evangelist for the cause. In 1912, a year before his death, Diesel predicted that vegetable oils would become as important a source of fuel as petroleum products.

This was, no doubt, a more appealing vision for the owners of peanut farms than for the owners of oil fields – and the impetus to make it happen largely dissipated with Rudolf Diesel's death. Hence the second conspiracy theory to inspire a speculatively sensationalist headline in a contemporary newspaper: 'Murdered by Agents from Big Oil Trusts'.

There's recently been a resurgence of interest in biodiesel. It's less polluting than oil fuel, but it's controversial – it competes for land with agriculture, pushing up food prices. In Rudolf Diesel's era, this was less of a concern: the population was much smaller, and the climate was more predictable. He was excited by the idea that his engine could help to develop poor, agricultural economies. How different might the world look today, if the most valuable land during the last hundred years wasn't where you could drill for oil, but where you could cultivate peanuts?

We can only guess – just as we'll never know for sure what happened to Rudolf Diesel. By the time his body bobbed up alongside another boat, ten days later, it was too badly decomposed for an autopsy; indeed, for the crew to be willing to take it on board at all. They extracted from Diesel's jacket his wallet, pocket knife and spectacles case, which his son later identified. The inventor's body was retaken by the waves.

32

Clocks

In 1845, a curious feature was added to the clock on St John's Church in Exeter, western England: another minute hand, running fourteen minutes faster than the original. This was, as *Trewman's Exeter Flying Post* explained, 'a matter of great public convenience', for it enabled the clock to exhibit, 'as well as the correct time at Exeter, the railway time'.

The human sense of time has always been defined by planetary motion. We talked of 'days' and 'years' long before we knew that the Earth rotates on its axis and orbits the sun; from the waxing and waning of the moon, we got the idea of a month. The sun's passage across the sky gives us phrases like 'mid-day', or 'high noon'. Exactly when the sun reaches its highest point depends, of course, on where you're looking from. If you happen to be in Exeter, you'll see it about fourteen minutes after someone in London.

Naturally, as clocks became commonplace, people set them by their local celestial observations. That was fine if you needed to coordinate only with other locals: if we both live in Exeter and say we'll meet at 7 p.m., it hardly matters that in London, 200 miles away, they think it's 7.14. But as soon as a train connects Exeter and London – stopping at multiple other towns, all with their own idea of what the time is – we face a logistical nightmare. Early train timetables valiantly informed travellers that 'London time is about four minutes earlier than Reading time, seven and a half minutes before Cirencester', and so on, but many understandably got hopelessly confused. More seriously, so did drivers and signalling staff, which raised the risk of collisions.

So the railways adopted 'railway time'. They based it on Greenwich Mean Time, set by the famous observatory in the London borough of Greenwich. Some municipal authorities quickly grasped the usefulness of standardising time across the country, and adjusted their own clocks accordingly. Others resented this high-handed metropolitan imposition, and clung to the idea that their time was – as the *Flying Post* put it, with charming parochialism – 'the correct time'. For years, the Dean of Exeter stubbornly refused to adjust the clock on the city's cathedral.

In fact, there's no such thing as 'the correct time'. Like the value of money, it's a convention that derives its usefulness from the widespread acceptance of others. But there is such a thing as accurate timekeeping. That dates from 1656, and a Dutchman named Christiaan Huygens.

There were clocks before Huygens, of course. Water clocks appear in civilisations from ancient Egypt to medieval Persia; others kept time from marks on candles. But even the most accurate devices might wander by fifteen minutes a day. This didn't matter much if you were a monk wanting to know when to pray, unless God is a stickler for punctuality. But there was one, increasingly important area of life where the inability to keep accurate time was of huge economic significance: sailing.

By observing the angle of the sun, sailors could figure

out their *latitude*: where you are from north to south. But their *longitude* – where you are from east to west – had to be guessed. Wrong guesses could, and frequently did, lead to ships hitting land hundreds of miles away from where navigators thought they were; sometimes literally hitting land, and sinking.

How could accurate timekeeping help? Remember why Exeter's clocks differed from London's, 200 miles away: high noon happened fourteen minutes later. If you knew when it was midday at London's Greenwich observatory – or any other reference point – you could observe the sun, calculate the time difference and work out the distance. Huygens' pendulum clock was sixty times more accurate than any previous device: but even fifteen seconds a day soon mounts up on long seafaring voyages, and pendulums don't swing neatly on the deck of a lurching ship.

Rulers of maritime nations were acutely aware of the longitude problem: the King of Spain offered a prize for solving it nearly a century before Huygens' work. Famously, it was a subsequent prize offered by the British government that led to a sufficiently accurate device being painstakingly refined, in the 1700s, by an Englishman named John Harrison.* It kept time to within a couple of seconds a day.

Since Huygens and Harrison, clocks have become much more accurate still. And since the Dean of Exeter's intransigence, the whole world has agreed on what to regard as 'the correct time' – coordinated universal time, or UTC,

---

* John Harrison solved the longitude problem, but he was never actually paid the prize he thought he deserved. In *Longitude* (1995) Dava Sobel makes a persuasive case that Harrison was unfairly deprived of his prize by the machinations of envious astronomers. But there's an alternative view: that because Harrison jealously guarded the details of how his clock worked, he did not provide a practical solution to the longitude problem – he merely demonstrated that he had such a solution within his gift.

as mediated by various global time zones which maintain the convention of 12 o'clock being at least vaguely near the sun's highest point. UTC is based on atomic clocks, which measure oscillations in the energy levels of electrons. The Master Clock itself, operated by the US Naval Observatory in north-west Washington DC, is actually a combination of several different clocks – the most advanced of which are four atomic fountain clocks, in which frozen atoms are launched into the air and cascade down again. If something goes wrong – and even a technician entering the room will alter the temperature and possibly the timing – then there are several back-up clocks, ready to take over at any nanosecond. The output of all this sophistication is accurate to within a second every three hundred million years.

Is there a point to such accuracy? We don't plan our morning commutes to the millisecond. In truth, an accurate wristwatch has always been more about prestige than practicality. For over a century, before the hourly beeps of early radio broadcasts, members of the Belville family made a living in London by setting their watches in Greenwich every morning and selling the time around the city, for a modest fee. Their clients were mostly tradesfolk in the horology business, for whom aligning their wares with Greenwich was a matter of professional pride.

But there are places where milliseconds now matter. One is the stock market: fortunes can be won by exploiting an arbitrage opportunity an instant before your competitors. Some financiers recently calculated it was worth spending 300 million dollars on drilling through mountains between Chicago and New York to lay fibre-optic cables in a slightly straighter line. That sped up communication between the two cities' exchanges by three milliseconds. One may reasonably wonder whether that's the most socially useful infrastructure the money could have bought, but the incentives for this kind of innovation are perfectly clear, and we can hardly be surprised if people respond to them.

The accurate keeping of universally accepted time also underpins computing and communications networks. But perhaps the most significant impact of the atomic clock – as it was first with ships, and then with trains – has been on travel.

Nobody now needs to navigate by the angle of the sun – we have GPS. The most basic of smartphones can locate you by picking up signals from a network of satellites: because we know where each of those satellites should be in the sky at any given moment, triangulating their signals can tell you where you are on earth. It's a technology that has revolutionised everything from sailing to aviation, surveying to hiking. But it works only if those satellites agree on the time.

GPS satellites typically house four atomic clocks, made from caesium or rubidium. Huygens and Harrison could only have dreamed of their precision, but it's still enough to misidentify your position by a couple of metres – a fuzziness amplified by interference as signals pass through the Earth's ionosphere. That's why self-driving cars need sensors as well as GPS: on the highway, a couple of metres is the difference between lane discipline and a head-on collision.

Meanwhile, clocks continue to advance: scientists have recently developed one, based on an element called ytterbium, that won't have lost more than a hundredth of a second by the time the sun dies and swallows up the Earth, in about five billion years. How might this extra accuracy transform the economy between now and then? Only time will tell.

33

The Haber-Bosch Process

It was a marriage of brilliant scientific minds. Clara Immerwahr had just become the first woman in Germany to receive a doctorate in chemistry. That took perseverance. Women couldn't study at the University of Breslau, so she asked each lecturer, individually, for permission to observe their lessons as a guest. Then she pestered to be allowed to sit the exam. The dean, awarding her doctorate, said 'Science welcomes each person, irrespective of sex'; he then undermined this noble sentiment by observing that a woman's duty was family, and he hoped this wasn't the dawn of a new era.

Clara saw no reason why getting married should interfere with her career. She was disappointed. Her husband turned out to be more interested in a dinner party hostess than a professional equal. She gave some lectures, but soon became discouraged when she learned that everyone assumed her husband had written them for her. He worked, networked, travelled and philandered; she was left holding the baby. Reluctantly, resentfully, she let her professional ambitions slide.

We'll never know what Clara Immerwahr might have

achieved, had attitudes to gender been different in early-twentieth-century Germany. But we can guess what she wouldn't have done. She would not – as her husband did – have pioneered chemical weapons. To help Germany win the First World War, he enthusiastically advocated gassing Allied troops with chlorine. She accused him of barbarity. He accused her of treason. After the first, devastatingly effective, use of chlorine gas – at Ypres, in 1915 – he was made an army captain. She took his gun and killed herself.

Clara and Fritz Haber had been married for fourteen years. Eight years into that time, Haber made a breakthrough that some now consider to be the most significant invention of the twentieth century. Without it, close to half the world's population would not be alive today.

The Haber-Bosch process uses nitrogen from the air to make ammonia, which can then be used to make fertilisers. Plants need nitrogen: it's one of their basic requirements, along with potassium, phosphorus, water and sunlight. In a state of nature, plants grow, they die, the nitrogen they contain returns to the soil, and new plants use it to grow. Agriculture disrupts that cycle: we harvest the plants and eat them.

From the earliest days of agriculture, farmers discovered various ways to prevent yields from declining over time – as it happened, by restoring nitrogen to their fields. Manure has nitrogen. So does compost. The roots of legumes host bacteria that replenish the soil's nitrogen; that's why it helps to include peas or beans in crop rotation. But these techniques struggle to fully satisfy a plant's appetite for nitrogen; add more, and the plant grows better.

It was only in the nineteenth century that chemists discovered this – and the irony that 78 per cent of the air is nitrogen, but not in a form plants can use. In the air, nitrogen consists of two atoms locked tightly together. Plants need those atoms 'fixed', or compounded with some other element: ammonium oxalate, for example, as found in guano, also known as bird poo; or potassium nitrate, also known as saltpetre and a main ingredient of gunpowder. Reserves of both guano and saltpetre were found in South America, mined, shipped around the world and dug into soil. But by the century's end, experts were fretting about what would happen when these reserves ran out.

If only it were possible to convert nitrogen from the air into a form plants could use.

That's exactly what Fritz Haber worked out how to do. He was driven partly by curiosity, partly by the patriotism that was later to lead him down the path to chemical warfare, and partly by the promise of a lucrative contract from the chemical company BASF. That company's engineer, Carl Bosch, then managed to replicate Haber's process on an industrial scale. Both men later won Nobel Prizes – controversially, in Haber's case, as many by then considered him a war criminal.

The Haber-Bosch process is perhaps the most significant example of what economists call technological substitution: when we seem to have reached some basic physical limit, then find a workaround. For most of human history, if you wanted more food to support more people, then you needed more land. But the thing about land is, as Mark Twain once joked, that they're not making it any more. The Haber-Bosch process provided a substitute: instead of more land, make nitrogen fertiliser. It was like alchemy: 'Brot aus Luft', as Germans put it. 'Bread from air'.

Well: bread from air, and quite a lot of fossil fuels. First of all, you need natural gas as a source of hydrogen, the element to which nitrogen binds to form ammonia. Then you need energy to generate extreme heat and pressure.

Haber discovered that was necessary, with a catalyst, to break the bonds between air's nitrogen atoms and persuade them to bond with hydrogen instead. Imagine the heat of a wood-fired pizza oven, with the pressure you'd experience 2 kilometres under the sea. To create those conditions on a scale sufficient to produce 160 million tonnes of ammonia a year – the majority of which is used for fertiliser – the Haber-Bosch process today consumes more than 1 per cent of all the world's energy.

That's a lot of carbon emissions, and it's far from the only ecological concern. Only some of the nitrogen in fertiliser makes its way via crops into human stomachs – perhaps as little as 15 per cent. Most of it ends up in the air or water. This is a problem for several reasons. Compounds like nitrous oxide are powerful greenhouse gases. They pollute drinking water. They create acid rain, which makes soils more acidic, which puts ecosystems out of kilter and biodiversity under threat. When nitrogen compounds run off into rivers, they likewise promote the growth of some organisms more than others; the results include ocean 'dead zones', where decaying algal blooms consume oxygen, suffocating marine life.

The Haber-Bosch process isn't the only cause of these problems, but it's a major one, and it's not going away: demand for fertiliser is projected to double in the coming century. In truth, scientists still don't fully understand the long-term impact on the environment of converting so much stable, inert nitrogen from the air into various other, highly reactive chemical compounds. We're in the middle of a global experiment.

One result of that experiment is already clear: plenty of food for lots more people. If you look at a graph of global population, you'll see it shoot upwards just as Haber-Bosch fertilisers start being widely applied. Again, Haber-Bosch wasn't the only reason for the spike in food yields; new varieties of crops like wheat and rice also played their part. Still, suppose we farmed with the best techniques available in Fritz Haber's time, the Earth would support about four billion people. The current population is around seven and a half billion, and although the growth rate has slowed, it has not stopped.

Back in 1909, as Fritz triumphantly demonstrated his ammonia process, Clara wondered whether the fruits of her husband's genius had been worth the sacrifice of her own. 'What Fritz has achieved in these eight years,' she wrote plaintively to a friend, 'I have lost.' She could hardly have imagined how transformative his work would be: on one side of the ledger, food to feed billions more human souls; on the other, a sustainability crisis that will need more genius to solve.

For Haber himself, the consequences of his work were not what he expected. As a young man, Haber had converted from Judaism to Christianity; he ached to be accepted as the German patriot he felt himself to be. Beyond his work on weaponising chlorine, the Haber-Bosch process also helped Germany in the First World War. Ammonia can make explosives, as well as fertiliser – not just bread from air, but bombs.

When the Nazis took power in the 1930s, however, none of these contributions outweighed his Jewish roots. Stripped of his job and kicked out of the country, Haber died, in a Swiss hotel, a broken man.

34

Radar

In Kenya's Rift Valley, Samson Kamau sat at home, wondering when he'd be able to get back to work. He should have been in a greenhouse on the shores of Lake Naivasha, as usual, packing roses for export to Europe. But the outbound cargo flights were grounded, because the Icelandic volcano Eyjafjallajökull had, without sparing the slightest thought for Samson, spewed a cloud of dangerous ash into Europe's airspace.

Nobody knew how long the disruption might last. Workers like Samson feared for their jobs; business owners had to throw away tons of flowers that were wilting in crates at Nairobi airport. As it happened, flights resumed within a few days. But the interruption dramatically illustrated just how much of the modern economy relies on flying, beyond the ten million passengers who get on flights every day. Eyjafjallajökull reduced global output in 2010 by nearly five billion dollars.

You could trace the extent of our reliance on air travel to many inventions. The jet engine, perhaps, or the aeroplane itself. But sometimes inventions need other inventions to unlock their full potential. For the aviation industry, that story starts with the invention of the death ray.

No, wait – it starts with an *attempt* to invent the death ray. This was back in 1935. Officials in the British Air Ministry were worried about falling behind Nazi Germany in the technological arms race. The death ray idea intrigued them: they'd been offering a thousand-pound prize for anyone who could zap a sheep at a hundred paces. So far, nobody had claimed it. But should they fund more active research? Was a death ray even possible? Unofficially, they sounded out Robert Watson Watt of the Radio Research Station. And he posed an abstract maths question to his colleague, Skip Wilkins.

Suppose, just suppose – said Watson Watt to Wilkins – that you had 8 pints of water, 1 kilometre above the ground. And suppose that water was at 98 degrees Fahrenheit, and you wanted to heat it to 105 degrees. How much radio frequency power would you require, from a distance of 5 kilometres?

Skip Wilkins was no fool. He knew that 8 pints was the amount of blood in an adult human, 98 degrees was normal body temperature and 105 degrees was warm enough to kill you, or at least make you pass out; which, if you're behind the controls of an aeroplane, amounts to much the same thing.

So Wilkins and Watson Watt understood each other, and they quickly agreed the death ray was hopeless: it would take too much power. But they also saw an opportunity. Clearly, the ministry had some cash to spend on research; perhaps Watson Watt and Wilkins could propose some alternative way for them to spend it?

Wilkins pondered: it might be possible, he suggested, to transmit radio waves and detect, from the echoes, the location of oncoming aircraft long before they could be seen. Watson Watt dashed off a memo to the Air Ministry's newly formed

Committee for the Scientific Survey of Air Defence: would they be interested in pursuing such an idea? They would indeed.

What Skip Wilkins was describing became known as radar. The Germans, the Japanese and the Americans all independently started work on it, too. But by 1940, it was the Brits who'd made a spectacular breakthrough: the resonant cavity magnetron, a radar transmitter far more powerful than its predecessors. Pounded by German bombers, Britain's factories would struggle to put the device into production. But America's factories could.

For months, British leaders plotted to use the magnetron as a bargaining chip for American secrets in other fields. Then Winston Churchill took power, and decided that desperate times called for desperate measures: Britain would simply tell the Americans what they had, and ask for help.

So it was that, in August 1940, a Welsh physicist named Eddie Bowen endured a nerve-racking journey with a black metal chest containing a dozen prototype magnetrons. First, he took a black cab across London: the cabbie refused to let the clunky metal chest inside, so Bowen had to hope it wouldn't fall off the roof rack. Then a long train ride to Liverpool, sharing a compartment with a mysterious, sharply dressed, military-looking man who spent the entire journey ignoring the young scientist and silently reading a newspaper. Then the ship across the Atlantic; what if it were hit by a German U-boat? The Nazis couldn't be allowed to recover the magnetrons; two holes were drilled in the crate to make sure it would sink if the boat did. But the boat didn't.

The magnetron stunned the Americans; their research was years off the pace. President Roosevelt approved funds for a new laboratory at MIT – uniquely, for the American war effort, administered not by the military but a civilian agency.

Industry got involved, and the very best American academics were headhunted to join Bowen and his British colleagues.

By any measure, Rad Lab was a resounding success. It spawned ten Nobel laureates. The radar it developed, detecting planes and submarines, helped to win the war. But urgency in times of war can quickly be lost in times of peace. It might have been obvious, if you thought about it, that civilian aviation needed radar, given how quickly it was expanding: in 1945, at the war's end, US domestic airlines carried seven million passengers; by 1955, it was thirty-eight million. And the busier the skies, the more useful radar would be at preventing collisions.

But rollout was slow and patchy. Some airports installed it; many didn't. In most airspace, planes weren't tracked at all. Pilots submitted their flight plans in advance, which should in theory ensure that no two planes were in the same place at the same time. But avoiding collisions ultimately came down to a four-word protocol: 'see and be seen'.

On the morning of 30 June 1956, two passenger flights departed from Los Angeles Airport, three minutes apart: one was bound for Kansas City, one for Chicago. Their planned flight paths intersected above the Grand Canyon, but at different heights. Then thunderclouds developed. One plane's captain radioed to ask permission to fly above the storm. The air traffic controller cleared him to go to '1000 on top' – a thousand feet above cloud cover. See and be seen.

Nobody knows for sure what happened: planes then had no black boxes, and there were no survivors. At just before 10.31 hours, air traffic control heard a garbled radio transmission: 'pull up!'; 'we are going in ...' From the pattern of the wreckage, strewn for miles across the canyon floor, the planes seem to have approached each other at a 25-degree angle, presumably through a cloud. Investigators speculated that both pilots were distracted by trying to find gaps in the clouds, so passengers could enjoy the scenery.

Accidents happen. The question is what risks we're willing to run for the economic benefits. That question's becoming pertinent again with respect to crowded skies: many people have high hopes for unmanned aerial vehicles, or drones. They're already being used for everything from movie-making to crop-spraying; companies like Amazon expect the skies of our cities soon to be buzzing with grocery deliveries. Civil aviation authorities are grappling with what to approve. Drones have 'sense-and-avoid' technology, and it's pretty good – but is it good enough?

The crash over the Grand Canyon certainly concentrated minds. If technology existed to prevent things like this, shouldn't we make more effort to use it? Within two years, what's now known as the Federal Aviation Administration was born in the United States. And today, American skies are about twenty times busier still. The world's biggest airports now see planes taking off and landing at an average of nearly two a minute. Collisions are absurdly rare, no matter how cloudy the conditions. That's thanks to many things, but it's largely thanks to radar.

35

Batteries

M urderers in early-nineteenth-century London sometimes tried to kill themselves before they were hanged.
Failing that, they asked friends to give their legs a good,
hard tug as they dangled from the gallows. They wanted to
make absolutely certain they'd be dead. Their freshly hanged
bodies, they knew, would be handed to scientists for ana-
tomical studies. They didn't want to survive the hanging and
regain consciousness while being dissected.

If George Foster, executed in 1803, had woken up on the
lab table, it would have been in particularly undignified circumstances. In front of an enthralled and slightly horrified
London crowd, an Italian scientist with a flair for showman-
ship was sticking an electrode up Foster's rectum.

Some in the audience thought Foster *was* waking up. The
electrically charged probe caused his lifeless body to flinch
and his fist to clench. Applied to his face, electrodes made his
mouth grimace and an eye twitch open. One onlooker was
apparently so shocked, he dropped dead shortly after. The
scientist had modestly assured his audience that he wasn't
actually intending to bring Foster back to life, but – well,

these were new and little-tested techniques. Who knew what might happen? The police were on hand, just in case Foster needed hanging again.

Foster's body was being *galvanised* – a word coined for Luigi Galvani, the Italian scientist's uncle. In 1780s Italy, Galvani had discovered that touching the severed legs of a dead frog with two different types of metal caused the legs to jerk. Galvani thought he had discovered 'animal electricity', and his nephew was carrying on the investigations. Galvanism briefly fascinated the public, inspiring Mary Shelley to write her story of Frankenstein.*

Galvani was wrong. There is no animal electricity. You can't bring hanged bodies back to life, and Victor Frankenstein's monster remains safely in the realms of fiction.

But Galvani was wrong in a useful way, because he showed his experiments to his friend and colleague Alessandro Volta, who had a better intuition about what was going on. The important thing, Volta realised, wasn't that the frog flesh was of animal origin; it was that it contained fluids that conducted electricity, allowing a charge to pass between the different types of metal. When the two metals were connected – Galvani's scalpel touching the brass hook on which the legs were hung – the circuit was complete, and a chemical reaction caused electrons to flow.

Volta experimented with different combinations of metal, and different substitutes for frogs' legs. In 1800, he showed

* Shelley thought of the idea during the 'year without a summer', the apocalyptic conditions in Europe that followed the eruption of Mount Tambora. Incessant rain confined Shelley and her clique – including Percy Shelley and Lord Byron – to a villa overlooking Lake Geneva. They competed to produce the most frightening story. As well as being influenced by Galvanism, Shelley's vision of a monster who was a homeless, friendless outcast may have echoed her experience of seeing starving peasants roaming from village to village in search of food. The same grim exposure to this suffering inspired the young Justus von Liebig to a life dedicated to preventing hunger.

that you could generate a constant, steady current by piling up sheets of zinc, copper and brine-soaked cardboard. Volta had invented the battery.

Like his friend Galvani, Volta gave us a word: *volt*. He also gave us an invention you might be using right now if you're listening to the audiobook, or reading on a tablet. Such portable devices are possible only thanks to the battery. Imagine, for a moment, a world without batteries: we'd be hand-cranking our cars, and getting tangled up in wires from our television remote controls.

Volta's insight won him admirers – indeed, Napoleon made him a count. But Volta's battery wasn't especially practical, not at first. The metals corroded, the salt water spilled, the current was short-lived, and it couldn't be recharged. It was 1859 before we got the first rechargeable battery, made from lead, lead dioxide and sulphuric acid. It was bulky and heavy, and acid sloshed out if you tipped it over. But it was useful – the same basic design still starts our cars. The first 'dry' cells, the familiar modern battery, came in 1886. The next big breakthrough took another century. It arrived in Japan.

In 1985, Akira Yoshino patented the lithium-ion battery; Sony later commercialised it. Researchers had been keen to make lithium work in a battery, as it's very light and highly reactive – lithium-ion batteries can pack large amounts of power into a small space. Unfortunately, lithium also has an alarming tendency to explode when exposed to air and water, so it took some clever chemistry to make it acceptably stable.

Without the lithium-ion battery, mobiles would likely have been much slower to catch on. Consider what cutting-edge battery technology looked like when Yoshino filed his patent. Motorola had just launched the world's first mobile phone, the DynaTAC 8000x: it weighed nearly a kilogram, and early adopters affectionately knew it as 'the brick'. Its talk time was thirty minutes.

The technology behind lithium-ion batteries has certainly improved: 1990s laptops were clunky and discharged rapidly; today's sleek ultraportables will last for a long-haul flight. Still, battery life has improved at a much slower rate than other laptop components, such as memory and processing power. Where's the battery that's light and cheap, recharges in seconds and never deteriorates with repeated use? We're still waiting.

Another major breakthrough in battery chemistry may be just around the corner. Or it may not. There's no shortage of researchers who hope they're onto the next big idea: some are developing 'flow' batteries, which work by pumping charged liquid electrolytes; some are experimenting with new materials to combine with lithium, including sulphur and air; some are using nanotechnology in the wires of electrodes to make batteries last longer. But history counsels caution: gamechangers haven't come along often.

Anyway, in the coming decades, the truly revolutionary development in batteries may not be in the technology itself, but in its uses. We're used to thinking of batteries as things that allow us to disconnect from the grid. We may soon see them as the thing that makes the grid work better.

Gradually, the cost of renewable energy is coming down. But even cheap renewables pose a problem: they don't generate power all the time. Even if the weather were perfectly predictable, you'd still have a glut of solar power on summer days and none on winter evenings. When the sun isn't shining and the wind isn't blowing, you need coal or gas or nuclear to keep the lights on – and once you've built those plants, why not run them all the time? A recent study of southeastern Arizona's grid weighed the costs of power cuts against the costs of carbon dioxide emissions, and concluded that solar should provide just 20 per cent of power. And Arizona is a pretty sunny place.

For grids to make more use of renewables means finding better ways of storing energy. One time-honoured solution is pumping water uphill when you have energy to spare, and then – when you need more – letting it flow back down through a hydropower plant. But that requires conveniently contoured mountainous terrain, and that's in limited supply. Could batteries be the solution?

Perhaps: it depends partly on the extent to which regulators nudge the industry in that direction, and partly also on how quickly battery costs come down.

Elon Musk hopes they'll come down very quickly indeed. The entrepreneur behind electric car-maker Tesla is building a gigantic lithium-ion battery factory in Nevada. Musk claims it will be the second-largest building in the world, next only to the one where Boeing manufacture their 747s. Tesla is betting that it can significantly wrestle down the costs of lithium-ion production, not through technological breakthroughs, but through sheer economies of scale.

Tesla needs the batteries for its vehicles, of course. But it's also among the companies already offering battery packs to homes and businesses: if you have solar panels on your roof, a battery in your house gives you the option of storing your surplus day-time energy for night-time use, rather than selling it back to the grid.

We're still a long way from a world in which electricity grids and transport networks can operate entirely on renewables and batteries. But the aim is becoming conceivable – and in the race to limit climate change, the world needs something to galvanise it into action. The biggest impact of Alessandro Volta's invention may be only just beginning.

36

Plastic

'Unless I am very much mistaken this invention will prove important in the future.' Leo Baekeland wrote those words in his journal on 11 July 1907. He was in a good mood, and why not? At forty-three years old, Baekeland had done well for himself.

He was born in Belgium. If it'd been up to his father, he'd still have been there, mending shoes. The father was a cobbler: he'd had no education, and he didn't understand why young Leo wanted one. He apprenticed the boy into the trade, aged just thirteen.

But his mother, a domestic servant, had other ideas. With her encouragement, Leo went to night school, and won a scholarship to the University of Ghent; by the age of twenty, he had a doctorate in chemistry. He married his tutor's daughter, they moved to New York, and Leo invented a new kind of photographic printing paper that made him a fortune – enough, at least, that he need never work again. He bought a house overlooking the Hudson River in Yonkers, just north of New York City. And he built a home laboratory, to indulge his love of tinkering with chemicals. In July 1907, he was experimenting with formaldehyde and phenol.

The cheerful journal entries continued. 18 July: 'Another hot sultry day. But I do not mind it and thoroughly appreciate the luxury of being allowed to stay home in shirt sleeves and without a collar.' Not all rich men were so happy, Baekeland knew: 'How about these Slave millionaires in wall street [sic] who have to go to their money making pursuit notwithstanding the sweltering heat. All day spent in laboratory', he concluded with an unmistakeable note of satisfaction. Perhaps he mused about whom he had to thank for this enjoyable, carefree life – the next day's journal entry records that he wired a hundred dollars to his mother. Four days later: 'This is the 23rd anniversary of my Doctorship ... How these twenty three years have gone fast ... Now I am again a student, and a student I will remain until death calls me again to rest.'

Baekeland wasn't entirely right about that. By the time death called him, at the age of eighty, his mental health had declined; he'd become an increasingly eccentric recluse, living off tinned food in his Florida mansion. But what a life he lived in the meantime. He made a second fortune. He became famous enough that *Time* magazine put his face on the cover without needing to mention his name – just the words: 'It will not burn. It will not melt'.

What Leo Baekeland invented that July was the first fully synthetic plastic. He called it Bakelite.

And he was right about the future importance. Plastics now are everywhere. When the author Susan Freinkel set out to write a book on them, she spent a day noting down everything she touched that was plastic: the light switch, the toilet seat, the toothbrush, the toothpaste tube; she also noted everything that wasn't – the toilet paper, the wooden floor, the porcelain tap. By the day's end, she'd listed 102 items that weren't made of plastic – and 196 that were. The world makes so much plastic, it takes about 8 per cent of oil production – half for raw material, half for energy.

The Bakelite Corporation didn't hold back in its advertising blurb: humans, it said, had transcended the old taxonomy of animal, mineral and vegetable; now we had a 'fourth kingdom, whose boundaries are unlimited'. That sounds hyperbolic, but it was true. Scientists previously had thought about improving or mimicking natural substances: earlier plastics, like celluloid, were based on plants, and Baekeland himself had been seeking an alternative to shellac, a resin secreted by beetles that was used for electrical insulation. Yet he quickly realised that Bakelite could become far more versatile than that. The Bakelite Corporation christened it 'The Material of a Thousand Uses', and, again, it wasn't far wrong: Bakelite went into telephones and radios, guns and coffee pots, billiard balls and jewellery. It was even used in the first atomic bomb.

Bakelite's success shifted mindsets. Bakelite – as *Time* would eventually celebrate – did not burn or melt and it was a good insulator. It looked good and it was cheap. So what other artificial materials might be possible, lighter or stronger or more flexible than you might find in nature, yet for a bargain price? In the 1920s and 1930s, plastics poured out of labs around the world. There was polystyrene, often used for packaging; nylon, popularised by stockings; polyethylene, the stuff of plastic bags. As the Second World War stretched natural resources, production of plastics ramped up to fill the gap. And when the war ended, exciting new products like Tupperware hit the consumer market.

But they weren't exciting for long: the image of plastic gradually changed. In 1967, the movie *The Graduate* famously started with the central character, Benjamin Braddock, receiving unsolicited career advice from a self-satisfied older neighbour. 'Just one word,' the neighbour promises, steering Benjamin towards a quiet corner, as if about to reveal the secret to life itself. 'Plastics!' The line became much-quoted, because it crystallised the changing connotations of the word: for the older neighbour's generation, 'plastic' still meant opportunity and modernity; for the likes of young Benjamin, it stood for all that was phoney, superficial, ersatz.

Still: it was great advice. Half a century on, despite its image problem, plastic production has grown about twenty-fold. It'll double again in the next twenty years. That's also despite growing evidence of environmental problems. Some of the chemicals in plastics are thought to affect how animals develop and reproduce. When plastics end up in landfill, those chemicals can eventually seep into groundwater; when they find their way into oceans, some creatures eat them. One estimate is that by 2050, all the plastic in the sea will weigh more than all the fish. (It's not clear how confident we can be of this claim, since nobody has managed to weigh either quantity.)

And there's another side to the ledger – plastic has benefits that aren't just economic, but environmental too. Vehicles made with plastic parts are lighter, and so use less fuel. Plastic packaging keeps food fresh for longer, and so reduces waste. If bottles weren't made of plastic, they'd be made of glass; which would you rather gets dropped in your children's playground?

Eventually, we'll have to get better at recycling plastic, if only because oil won't last forever. Some plastics can't be recycled; Bakelite is one. Many more could be, but aren't. Only about a seventh of plastic packaging is recycled – far less than for paper or steel; for other products, that rate is lower still. Improving it will take effort from everyone. You may have seen little triangles on plastic, with numbers from one to seven. They're called Resin Identification Codes, and they're one initiative of the industry's trade association. They help with recycling, but the system's far from perfect. If the industry could do more, then so could many governments: recycling rates differ hugely around the world. One success story is Taipei: it's changed its culture of waste by making it easy for citizens to recycle, and even fining them if they don't.

How about technological solutions? Fans of science fiction will enjoy one recent invention, the ProtoCycler: feed it your plastic waste, and it gives you filament for your 3D printer. Unfortunately, just like corrugated cardboard, plastic cannot be recycled indefinitely before the quality becomes unacceptable. Still – the ProtoCycler is as close as we can get today to *Star Trek*'s replicator.

In its day, Bakelite must have felt as revolutionary as a *Star Trek* replicator feels to us. Here was a simple, cheap synthetic product that was tough enough to replace ceramic tableware or metal letter openers, yet beautiful enough to be used as jewellery, and could even replace precious ivory. It was a miracle material, even though – like all plastics today – we now take it for granted.

But manufacturers today haven't given up on the idea that you can make something precious and practical from something cheap and worthless. The latest techniques 'upcycle' plastic trash. One, for example, turns old plastic bottles into a material resembling carbon fibre – it may be strong and light enough to make recyclable airplane wings. In general, mixing discarded plastic with other waste materials – and a dash of nanoparticles – promises to create new materials, with new properties.

Leo Baekeland would have approved.

VI

THE VISIBLE HAND

Adam Smith's 'invisible hand' is the most famous metaphor in economics. He used the phrase three times, most famously in the *Wealth of Nations* in 1776, when he wrote that as each individual tries to invest, 'he intends only his security . . . he intends only his own gain, and he is in this as in many other cases, led by an invisible hand to promote an end which was no part of his intention'.

Exactly what Smith meant by the invisible hand is excitedly debated by scholars to this day. But for modern economists, the metaphor has taken on a life well beyond Smith's intentions. It now describes the idea that when individuals and companies compete in the marketplace, the outcome is socially beneficial: products are produced efficiently, and they're consumed by the people who value them most. Perhaps there are a few free-market fans who think this is an accurate description of how markets actually work, but the mainstream of the economics profession views it more as a useful starting point. Markets do tend to allocate resources well, but that tendency isn't a guarantee. The invisible hand does not always guide us: sometimes we need the visible hand of government too.

We've seen many examples of this. Radar is now an

indispensable civilian technology, but it was developed for military purposes and generously funded by governments. The iPhone is a work of capitalist genius, by some measures the most successful product that has ever been produced – but it rested on government funding of computing, the internet, GPS and the World Wide Web.

Some of the most important inventions that shaped the modern economy weren't just helped along the way by government, they were entirely the creation of the state – for example the limited liability company, intellectual property, and most obviously the welfare state itself.

If markets can fail, though, so too can government regulators. Japanese women were denied access to the contraceptive pill for decades because regulators refused to approve it. One of the major obstacles that Malcom McLean faced in introducing container shipping was the bureaucracy of the US freight regulators, who seemed to feel that the only acceptable option was that nothing should ever change. And when researchers developed public key cryptography, the remarkable technology that makes internet commerce possible, the US government tried to shut them down.

Sometimes the state provides the foundations for new ideas; sometimes it's the chief obstacle. And sometimes it's more complicated than that – as we'll see with the case of M-Pesa, an idea that depended on seed funding from the British government and benign neglect from the Kenyan authorities. The dance between the state and the market continues to fascinate. Sometimes the state steps in, sometimes the state steps back, and sometimes the state simply tramples on everyone's toes.

37

The Bank

On London's busy Fleet Street, just opposite Chancery Lane, there is a stone arch through which anyone may step, and travel back in time. Just a few yards south, in a quiet courtyard, there is a strange, circular chapel – and next to it, on a column, is a statue of two knights sharing a single horse. The chapel is Temple Church, consecrated in 1185 as the London home of the Knights Templar, a religious order. But Temple Church is not just an important architectural, historical and religious site. It is also London's first bank.

The Knights Templar were warrior monks: they were a religious order, with a theologically inspired hierarchy, mission statement and code of ethics, but they were also heavily armed and dedicated to a holy war. How did those guys get into the banking game?

The Templars dedicated themselves to the defence of Christian pilgrims to Jerusalem. The city had been captured by the first Crusade in 1099 and pilgrims began to stream in, travelling thousands of miles across Europe. And if you're a pilgrim, you have a problem: you need to somehow fund months of food and transport and accommodation and yet you also want to avoid carrying huge sums of cash around, because that makes you a target for robbers. Fortunately, the Templars had that covered. A pilgrim could leave his cash at Temple Church in London, and withdraw it in Jerusalem. Instead of carrying cash, he'd carry a letter of credit. The Knights Templar were the Western Union of the Crusades.

We don't actually know how the Templars made this system work and protected themselves against fraud. Was there a secret code verifying the document and the identity of the traveller? We can only guess. But that wouldn't be the only mystery to shroud the Templars, an organisation sufficiently steeped in legend that Dan Brown set a scene of *The Da Vinci Code* in Temple Church.

Nor were the Templars the first organisation in the world to provide such a service. Several centuries earlier, Tang dynasty China used 'feiqian' – flying money, a two-part document allowing merchants to deposit profits in a regional office, and reclaim their cash back in the capital. But that system was operated by the government. Templars were much closer to a private bank – albeit a private bank owned by the Pope, allied to kings and princes across Europe, and run by a partnership of monks sworn to poverty.

The Knights Templar did much more than transferring money across long distances. They provided a range of recognisably modern financial services. If you wanted to buy a nice island off the west coast of France – something King Henry III of England did in the 1200s with the island of Oléron, north west of Bordeaux – the Templars could broker the deal. The King paid two hundred pounds a year for five years to the Temple in London, then when his men took possession of the island, the Templars made sure that the island's previous owner was paid. Oh, and the Crown Jewels of England, stored today at the Tower of London? In the 1200s, the Crown

Jewels were at the Temple – security on a loan. That was the Templars operating as a very high-end pawn broker.

The Knights Templar weren't Europe's bank forever, of course. The order lost its reason to exist after European Christians completely lost control of Jerusalem in 1244; the Templars were eventually disbanded in 1312. But then who was to step into the banking vacuum?

If you had been at the great fair of Lyon in 1555, you could have seen the answer. Lyon's fair was the greatest market for international trade in all Europe and dated back to Roman times. But at this particular fair, gossip was starting to spread. There was this Italian merchant, see him? He was making a fortune. But how? He bought nothing and had nothing to sell. All he had was a desk and an inkstand. And he sat there, day after day as the fair continued, receiving other merchants and signing their pieces of paper, and somehow becoming very rich. Extraordinary. And frankly, to the locals, very suspect.

But to a new international elite of Europe's great merchant houses, this particular Italian's activities were perfectly legitimate. He had a very important role: he was buying and selling debt, and in doing so he was creating enormous economic value.

Here is how the system worked. A merchant from Lyon who wanted to buy – say – Florentine wool could go to this banker and borrow something called a bill of exchange. The bill of exchange was a credit note, an IOU. This IOU wasn't denominated in the French livre or Florentine lira. Its value was expressed in the *écu de marc*, a private currency used by this international network of bankers. And if the Lyonnais merchant travelled to Florence – or sent his agents there – the bill of exchange from the banker back in Lyon would be recognised by bankers in Florence, who would gladly exchange it for local currency.

Through this network of bankers, then, a local merchant could not only exchange currencies but also exchange his creditworthiness in Lyon for creditworthiness in Florence, a city where nobody had ever heard of him. That's a valuable service. No wonder that the mysterious banker was rich. And every few months, agents of this network of bankers would meet at the great fairs like Lyon's, go through their books, net off all the credit notes against each other and settle their remaining debts.

Our financial system today still has a lot in common with this system. An Australian with a credit card can walk into a supermarket in – well, let's say Lyon, why not? – and she can walk out with groceries. The supermarket checks with a French bank, the French bank talks to an Australian bank, and an Australian bank approves the payment, happy that this woman is good for the money.

But this web of banking services has always had a darker side to it. By turning personal obligations into internationally tradeable debts, these medieval bankers were creating their own private money – and that private money was outside the control of Europe's kings. They were rich and powerful, and they didn't need the coins minted by the sovereign.

That description rings true even today. International banks are locked together in a web of mutual obligations that defies understanding or control. They can use their international reach to try to sidestep taxes and regulations. And, since their debts to each other are a very real kind of private money, when the banks are fragile, the entire monetary system of the world also becomes fragile.

We're still trying to figure out what to do with these banks. We can't live without them, it seems, and yet we're not sure we want to live with them. Governments keep searching for ways to hold them in check. Sometimes the approach has been *laissez-faire*. Sometimes not.

Few regulators have been quite as ardent as King Philip IV of France. He owed money to the Templars, and they refused to forgive his debts. So in 1307, on the site of what is now the Temple stop on the Paris Metro, King Philip launched a raid on the Paris Temple – the first of a series of attacks across Europe. Templars were tortured and forced to confess any sin the Inquisition could imagine. The order of the Templars was disbanded by the Pope. The London Temple was rented out to lawyers. And the last grandmaster of the Templars, Jacques de Molay, was brought to the centre of Paris and publicly burned to death.

38

Razors and Blades

'There are clouds upon the horizon of thought, and the very air we breathe is pregnant with life that foretells the birth of a wonderful change.' So begins a book written in 1894, by a man who had a vision that has ended up shaping how the modern economy works.

The book argues that 'our present system of competition' breeds 'extravagance, poverty, and crime'. It advocates a new system of 'equality, virtue, and happiness', in which just one corporation – the United Company – will make all of life's necessities, as cost-effectively as possible. These, by the way, are 'food, clothing, and habitation'. Industries which 'do not contribute' to life's necessities will be destroyed. Sorry, bankers and lawyers. That means you.

It's the end of money, too: instead, the manual labour required to make life's necessities will be shared out 'with perfect justice'. It should take only about five years of each person's life. The rest will be freed up for intellectual pursuits: ambitious people will compete not for material wealth, but to win recognition for promoting the 'welfare and happiness' of their fellow beings.

The plan gets more specific. All of this will take place in a city called Metropolis, located between Lake Erie and Lake Ontario, where Canada meets New York State. Metropolis will run on hydropower. It'll be the only city in North America. Its citizens will live in 'mammoth apartment houses ... upon a scale of magnificence such as no civilization has ever known'. These buildings will be circular, 600 feet across, and separated by twice that distance of 'avenues, walks, and gardens'. Artificial parks will feature 'pillars of porcelain tile' with 'domes of colored glass in beautiful designs'. They'll be an 'endless gallery of loveliness'.

I said that the author of this florid utopia had a vision that's ended up shaping the economy. As you may have guessed, it wasn't this particular vision. No, it was another idea, which he had a year later. His name was King Camp Gillette, and he invented the disposable razor blade.

You may be wondering why that was so significant. Here's one illustration: if you've ever bought replacement cartridges for an inkjet printer, you are likely to have been annoyed to discover that they cost almost as much as you paid for the printer itself. That seems to make no sense. The printer's a reasonably large and complicated piece of technology. How can it possibly add only a negligible amount to the cost of supplying a bit of ink in tiny plastic pots?

The answer, of course, is that it doesn't. But for a manufacturer, selling the printer cheaply and the ink expensively is a business model that makes sense. After all, what's your alternative: buy a whole new printer from a rival manufacturer? As long as that's even slightly more expensive than the new ink for your current printer, you'll reluctantly pay up.

That business model is known as two-part pricing. It's also known as the 'razor and blades' model, because that's where it first drew attention – sucker people in with an attractively priced razor, then repeatedly fleece them for extortionately priced replacement blades.

King Camp Gillette invented the blades that made it possible. Before this, razors were bigger, chunkier affairs – and a significant enough expense that when the blade got dull, you'd sharpen (or 'strop') it, not chuck it away and buy another. Gillette realised that if he devised a clever holder for the blade, to keep it rigid, he could make the blade much thinner – and cheaper to produce.

He didn't immediately hit upon the two-part pricing model, though: initially, he made *both* parts expensive. Gillette's razor cost five dollars – about a third of the average worker's weekly wage; his philosophical concerns about 'extravagance' and 'poverty' seem not to have clouded his business decisions. So eye-wateringly exorbitant was the Gillette razor that the 1913 Sears catalogue offered it with an apology that Sears wasn't legally allowed to discount the price, along with an annoyed-sounding disclaimer: 'Gillette Safety Razors are quoted for the accommodation of some of our customers who want this particular razor. We don't claim that this razor will give better satisfaction than the lower priced safety razors quoted on this page.'

The model of cheap razors and expensive blades evolved only later, as Gillette's patents expired and competitors got in on the act. Nowadays, two-part pricing is everywhere. Consider the PlayStation 4. Every time Sony sells one, it loses money: the retail price is less than it costs to manufacture and distribute. But that's okay, because Sony coins it in whenever a PlayStation 4 owner buys a game. Or how about Nespresso? Nestlé makes its money not from the machine, but the coffee pods.

Obviously, for this model to work you need some way to prevent customers putting cheap, generic blades in your razor.

One solution is legal: patent-protect your blades. But patents don't last forever. Patents on coffee pods have started expiring, so brands like Nespresso now face competitors selling cheap, compatible alternatives. Some are looking for another kind of solution: technological. Just as other people's games don't work on the PlayStation, some coffee companies have put chip readers in their machines to stop you sneakily trying to brew a generic cup of joe.

Two-part pricing models work by imposing what economists call 'switching costs'. Want to brew another brand's coffee? Then buy another machine. They're especially prevalent with digital goods. If you have a huge library of games for your PlayStation, or books for your Kindle, it's a big thing to switch to another platform.

Switching costs don't have to be financial. They can come in the form of time, or hassle. Say I'm already familiar with Photoshop; I might prefer to pay for an expensive upgrade than buy a cheaper alternative, which I'd then have to learn how to use. That's why software vendors offer free trials. It's also why banks and utilities offer special 'teaser' rates to draw people in: when they quietly raise the price, many won't bother to change.

Switching costs can be psychological, too – a result of brand loyalty. If the Gillette company's marketing department persuades me that generic blades give an inferior shave, then I'll happily keep paying extra for Gillette-branded blades. That may explain the otherwise curious fact that Gillette's profits *increased* after his patents expired and competitors could make compatible blades. Perhaps, by then, customers had got used to thinking of Gillette as a high-end brand.

But the two-part pricing model pioneered by Gillette is highly inefficient, and economists have puzzled over why consumers stand for it. The most plausible explanation is that they get confused by the two-part pricing. Either they don't realise they'll be exploited later, or they do realise but find it hard to pick the best deal out of a confusing menu of options. It's just the kind of situation where you'd expect government regulators to step in and enforce clarity, as they have in many other situations where advertised prices can mislead: when there are compulsory added costs, for example, or spurious claims of reductions from a previously higher price.

And regulators around the world have tried to find rules that will prevent this sort of confusopoly, but finding rules that work has proved difficult. Perhaps that's not surprising, as two-part pricing isn't always a cynical bait-and-switch – often, it's a perfectly reasonable and efficient way for a company to cover its costs. For example, an electricity utility might charge a lump sum for maintaining a connection to the grid, and then a low unit price per kilowatt hour supplied. But while such a pricing scheme makes perfect sense, it can still leave customers confused as to which deals are the best.

The irony is that the cynical razors-and-blades model – charging customers a premium for basics like ink and coffee – is about as far as you can get from King Camp Gillette's vision of a single, United Company producing life's necessities as cheaply as possible. In his book's peroration, Gillette reached new heights of purple prose: 'Come one, come all, and join the ranks of an overwhelming United People's Party . . . Let us tear asunder the chrysalis which binds within its folds the intellect of man, and let the polar star of every thought find its light in nature's truths.' Evidently, it's easier to inspire a new model for business than a new model for society.

39

Tax Havens

Would you like to pay less tax? One way is to make a sandwich: specifically, a 'double Irish with a Dutch sandwich'. Suppose you're American. You set up a company in Bermuda, and sell it your intellectual property; then it sets up a subsidiary in Ireland. Now set up *another* company in Ireland: it bills your European operations for amounts resembling their profits. Now start a company in the Netherlands. Have your second Irish company send money to your Dutch company, which immediately sends it back to your first Irish company. You know, the one headquartered in Bermuda.

Are you bored and confused yet? If so, that's part of the point. If two-part pricing sometimes confuses customers, it's a model of simplicity compared to cross-border tax laws. Tax havens depend on making it, at best, very difficult to get your head around financial flows, and at worst, impossible to find out any facts. Accounting techniques that make your brain hurt enable multinationals like Google, eBay and Ikea to minimise their tax bills – completely legally.

You can see why people get upset about this. Taxes are a bit like membership fees for a club: it feels unfair to dodge the fees, but still expect to benefit from the services the club provides its members – defence, police, roads, sewers, education and so on. But tax havens haven't always had such a bad image. Sometimes they've functioned like any other safe haven, allowing persecuted minorities to escape the oppressive rules of home. Jews in Nazi Germany, for example, were able to ask secretive Swiss bankers to hide their money. Unfortunately, secretive Swiss bankers soon undid the good this did their reputation by proving just as happy to help the Nazis hide the gold they managed to steal, and reluctant to give it back to the people they stole it from.

Nowadays, tax havens are controversial for two reasons: tax avoidance and tax evasion. Tax *avoidance* is legal. It's the stuff of double Irish, Dutch sandwiches. The laws apply to everyone: smaller businesses and even ordinary individuals could set up border-hopping legal structures, too. They just don't earn enough to justify the accountants' fees.

If everyday folk want to reduce their tax bill, their options are limited to various forms of tax *evasion*, which is illegal: VAT fraud, undeclared cash-in-hand work, or taking too many cigarettes through the 'nothing to declare' lane at customs. The British tax authorities reckon that much evaded tax comes from countless such infractions, often small-time stuff, rather than the wealthy entrusting their money to bankers who've shown they can keep a secret. But it's hard to be sure. If we could measure the problem exactly, it wouldn't exist in the first place.

Perhaps it's no surprise that banking secrecy seems to have started in Switzerland: the first known regulations limiting when bankers can share information about their clients were passed in 1713 by the Great Council of Geneva. Secretive Swiss banking really took off in the 1920s, as many European nations hiked taxes to repay their debts from the First World

War – and many rich Europeans looked for ways to hide their money. Recognising how much this was boosting their economy, in 1934 the Swiss doubled down on the credibility of their promise of banking secrecy: they made it a criminal offence for bankers to disclose financial information.

The euphemism for a tax haven these days, of course, is 'offshore' – and Switzerland doesn't even have a coastline. Gradually, tax havens have emerged on islands such as Jersey or Malta, or most famously, in the Caribbean. There's a logistical reason for this: a small island isn't much good for manufacturing or agriculture, so financial services are an obvious alternative. But the real explanation for the rise of the offshore haven is historical: the dismantling of European empires in the decades after the Second World War. Unwilling to prop up Bermuda or the British Virgin Islands with explicit subsidies, the United Kingdom instead encouraged them to develop financial expertise, plugged into the City of London. But the subsidy happened anyway – it was implicit, and perhaps accidental, but tax revenue steadily leaked away to these islands.

The economist Gabriel Zucman came up with an ingenious way to estimate the wealth hidden in the offshore banking system. In theory, if you add up the assets and liabilities reported by every global financial centre, the books should balance – but they don't. Each individual centre tends to report more liabilities than assets. Zucman crunched the numbers and found that, globally, total liabilities were 8 per cent higher than total assets. That suggests at least 8 per cent of the world's wealth is illegally unreported. Other methods have come up with even higher estimates.

The problem is particularly acute in developing countries. For example, Zucman finds 30 per cent of wealth in Africa is hidden offshore. He calculates an annual loss of fourteen billion dollars in tax revenue. That would build plenty of schools and hospitals.

Zucman's solution is transparency: creating a global register of who owns what, to end banking secrecy and anonymity-preserving shell corporations and trusts. That might well help with tax evasion. But tax avoidance is a subtler and more complex problem.

To see why, imagine I own a bakery in Belgium, a dairy in Denmark and a sandwich shop in Slovenia. I sell a cheese sandwich, making one euro of profit. On how much of that profit should I pay tax in Slovenia, where I sold the sandwich, or Denmark, where I made the cheese, or Belgium, where I baked the bread? There's no obvious answer. As rising taxes met increasing globalisation in the 1920s, the League of Nations devised protocols for handling such questions. They allow companies some leeway to choose where to book their profits. There's a case for that, but it opened the door to some dubious accounting tricks. One widely reported example may be apocryphal but it illustrates the logical extreme of these practices: a company in Trinidad sold ball-point pens to a sister company for 8500 dollars apiece. The result: more profit booked in low-tax Trinidad; less in higher-tax regimes elsewhere.

Most such tricks are less obvious, and consequently harder to quantify. Still, Zucman estimates that 55 per cent of US-based companies' profits are routed through some unlikely-looking jurisdiction such as Luxembourg or Bermuda, costing the US taxpayer 130 billion dollars a year. Another estimate puts the losses to developing country governments at many times the amount they get in foreign aid.

Solutions are conceivable: profits could be taxed globally, with national governments devising ways to apportion which profit is deemed taxable where. A similar formula already exists to apportion national profits made by US companies to individual states.

But that would need political desire to tackle tax havens. And while recent years have seen some initiatives, notably by the OECD, they've so far lacked teeth. Perhaps this shouldn't surprise us, given the incentives involved. Clever people can earn more from exploiting loopholes than trying to close them. Individual governments face incentives to compete to lower taxes, because a small percentage of something is better than a large percentage of nothing; for tiny, palm-fringed islands it can even make sense to set taxes at 0 per cent, as the local economy will be boosted by the resulting boom in law and accounting.

Perhaps the biggest problem is that tax havens mostly benefit financial elites, including some politicians and many of their donors. Meanwhile, pressure for action from voters is limited by the boring and confusing nature of the problem.

Sandwich, anyone?

40

Leaded Petrol

Leaded petrol was safe. Its inventor was sure of it. Facing sceptical reporters at a press conference, Thomas Midgley dramatically produced a container of tetraethyl lead – the additive in question – and proceeded to wash his hands in it. 'I'm not taking any chance whatever,' Midgley declared. 'Nor would I take any chance doing that every day.'

Midgley was, perhaps, being a little disingenuous. He might have mentioned that he'd recently spent several months in Florida, recuperating from lead poisoning.

Some of those who'd been making Midgley's invention hadn't been so lucky, and this was why the reporters were interested. One Thursday, in October 1924, at a Standard Oil plant in New Jersey, a worker named Ernest Oelgert had started hallucinating. By Friday, he was running around the laboratory, screaming in terror. On Saturday, with Oelgert dangerously unhinged, his sister called the police; he was taken to hospital and forcibly restrained. By Sunday, he was dead. Within a week, so were four of his laboratory work-mates – and thirty-five more were in hospital.

Only forty-nine people worked there.

None of this surprised workers elsewhere in Standard Oil's facility. They knew there was a problem with tetraethyl lead. They referred to the lab where it was developed as 'the loony gas building'. Nor should it have shocked Standard Oil, General Motors or the Du Pont corporation, the three companies involved with adding tetraethyl lead to gasoline. The first production line, in Ohio, had already been shut down after two deaths. A third plant, elsewhere in New Jersey, had also seen deaths; workers kept hallucinating insects and trying to swat them away – the lab was known as 'the house of butterflies'.

Leaded petrol is banned now, almost everywhere. It's among the many regulations that shape the modern economy. And yet 'regulation' has become a dirty word: politicians often promise to sweep them away; you rarely hear calls for more. It's a trade-off, between protecting people and imposing costs on business. And the invention of leaded petrol marked one of the first times that trade-off sparked a fierce public controversy.

Scientists were alarmed. Was it really sensible to add lead to petrol, when cars belch fumes onto city streets? Thomas Midgley breezily responded that 'the average street will probably be so free from lead that it will be impossible to detect it or its absorption'. Was his insouciance based on data? Well, no. Scientists urged the government to investigate, and they did – with funding provided by General Motors, on condition they got to approve the findings.

In the middle of the media frenzy over Ernest Oelgert's poisoned workmates, the report landed. It gave tetraethyl lead a clean bill of health. And the public greeted it with scepticism. Under pressure, the government organised a conference in Washington DC, in May 1925. It was a showdown. In one corner: Frank Howard, vice president of the

Ethyl Corporation – a joint venture of General Motors and Standard Oil. He called leaded petrol a 'gift of God', arguing that 'continued development of motor fuels is essential in our civilization'.

In the other corner: Dr Alice Hamilton, the country's foremost authority on lead. She argued that leaded petrol was a chance not worth taking: 'Where there is lead', she said, 'some case of lead poisoning sooner or later develops, even under the strictest supervision.'

Hamilton knew that lead had been poisoning people for thousands of years. In 1678, workers who made lead white – a pigment for paint – were described as suffering ailments including 'dizziness in the Head, with continuous great pain in the Brows, Blindness, Stupidity'. In Roman times, lead pipes carried water; the Latin for lead, *plumbum*, gave us the word plumber. Even then, some realised the folly. 'Water conducted through earthen pipes is more wholesome than that through lead,' wrote the civil engineer Vitruvius, two thousand years ago. 'This may be verified by observing the workers in lead, who are of a pallid colour.'

Eventually, the government decided to ignore Alice Hamilton, and Vitrivius. Leaded petrol got the go-ahead. Half a century later, they changed their minds. And a couple of decades after that, the economist Jessica Reyes noticed something interesting. Rates of violent crime were starting to go down. There are many reasons why this might have happened, but Reyes wondered: children's brains are especially susceptible to chronic lead poisoning. Is it possible that kids who didn't breathe leaded petrol fumes grew up to commit less violent crime?

Different US states had phased out leaded petrol at different times. So Reyes compared the dates of clean air legislation with subsequent crime data. She concluded that over half the drop in crime – 56 per cent – was because of cars switching to unleaded petrol.

That doesn't prove leaded petrol was wrong. When countries are poor, they might decide that pollution is a price worth paying for progress. Later, as their incomes grow, they decide they can afford to bring in laws that clean the environment up. Economists have a name for this pattern: it's called the 'environmental Kuznets curve'.

But for leaded petrol, it was never a trade-off worth making. True, the lead additive did solve a problem: it enabled engines to use higher compression ratios, which made cars more powerful. But ethyl alcohol had much the same effect, and it wouldn't mess with your head, unless you drank it. Why did General Motors push tetraethyl lead instead of ethyl alcohol? Cynics might point out that any old farmer could distil ethyl alcohol from grain – it couldn't be patented, or its distribution profitably controlled. Tetraethyl lead could.

There's another way to assess the economic benefit of leaded petrol: ask how much it cost to adapt cars to unleaded fuel, when clean air laws came in. Jessica Reyes crunched the numbers: it came to about twenty times less than the cost of all the crime. And that's before you count the other costs of kids breathing lead, like learning less in school.

How did the US get this so wrong for so long? It's the same story of disputed science and delayed regulation you could tell about asbestos, or tobacco, or many other products that slowly kill us. The government did, in the 1920s, call for continued research. And for the next four decades, those studies were conducted by scientists funded by the Ethyl Corporation and General Motors. It was only in the 1960s that American universities developed policies on conflicts of interest in research.

Today's economy isn't short of health scares. Is GM food safe? What about nanoparticles? Does wifi cause cancer? How do we tell the wise words of an Alice Hamilton from the ill-informed worries of an obstructive crank? We've learned something about research and regulation-making from disasters like leaded petrol, but you'd be optimistic to think the problem is solved completely.

And what of the scientist who first put lead in petrol? By all accounts, Thomas Midgley was a genial man; he may even have believed his own spin about the safety of a daily hand-wash in tetraethyl lead. But, as an inventor, his inspirations seem to have been cursed. His second major contribution to civilisation was the chlorofluorocarbon, or CFC – it improved refrigerators, but destroyed the ozone layer.

In middle age, afflicted by polio, Midgley applied his inventor's mind to lifting his weakened body out of bed. He devised an ingenious system of pulleys and strings. They tangled around his neck, and killed him.

41

Antibiotics in Farming

On a ramshackle pig farm near Wuxi, in Jiangsu province, China, a foreigner gets out of a taxi. The family that lives there is surprised: their little farm is at the end of a bumpy track through rice paddies. They don't get many foreigners turning up in taxis and asking to use the toilet.

The stranger's name is Philip Lymbery, and he runs a campaigning group called Compassion in World Farming. He's not here to berate the farmers about the living conditions of their pigs, although they're depressing. Sows are crammed into crates, with no room to move. The living conditions of the family aren't much happier: the toilet, Lymbery finds, is a hole in the ground between the house and the pig pen. No, Lymbery's here to investigate if pig manure is polluting the local waterways. He's tried to visit the large, commercial farms in the vicinity, but they don't want to see him. So he's turned up on spec at a family farm instead.

The farmer is happy to talk. Yes, they dump waste in the river. No, they're not supposed to. But that's okay – they just bribe the local official. Then Lymbery notices something. It's a pile of needles. He takes a closer look. They're antibiotics.

Have they been prescribed by a vet? No, the farmer explains. You don't need a prescription to buy antibiotics. And anyway, vets are expensive. Antibiotics are cheap. She injects her pigs with them routinely, and hopes that'll stop them getting sick.

She's far from alone. Cramped and dirty conditions on intensive farms are breeding grounds for disease, but routine, low doses of antibiotic can help to keep disease in check. Antibiotics also fatten animals. Scientists are studying gut microbes for clues as to why that is, but farmers don't need to know why: they simply know that they make more money from fatter animals. No wonder more antibiotics are injected into healthy animals than sick humans. In the big emerging economies, where demand for meat is growing as incomes rise, use of agricultural antibiotics is set to double in twenty years.

The widespread use of antibiotics where they're not really needed isn't restricted to agriculture. Many doctors are guilty, too – and they should know better. So should the regulators who allow people to buy antibiotics over the counter. But the bacteria don't care who is to blame. They busily evolve resistance to drugs, and public health experts fear we are entering a post-antibiotic age. One recent review estimated that drug-resistant bugs could kill ten million people a year by 2050 – more than currently die from cancer. It's hard to put a monetary value on antibiotics becoming useless, but the review tried. The figure it came up with: a hundred trillion dollars. You might think we'd be doing everything possible to preserve antibiotics' life-saving power. Sadly, you'd be wrong.

The story of antibiotics starts with a healthy dose of serendipity. A young man named Alexander Fleming was earning a wage through a boring job in shipping when his uncle died, leaving him enough money to quit and enrol at St Mary's Hospital Medical School in London instead. There, he became a valued member of the rifle club. The captain of the shooting team didn't want to lose Fleming when his studies were over, so he lined up a job for Fleming. That's how Fleming became a bacteriologist. Then in 1928 Fleming didn't bother to tidy up his petri dishes before going back home to Scotland on holiday. On his return, he famously noticed that one dish had become mouldy in his absence, and the mould was killing the bacteria he'd used the dish to cultivate.

Fleming tried to investigate further by making more mould, but he wasn't a chemist – he couldn't figure out how to make enough. He published his observations, but nobody paid attention. A decade passed, then more serendipity: in Oxford, Ernst Chain was flicking through back copies of medical journals when he chanced upon Fleming's old article. And Chain, a Jew who'd fled Nazi Germany, *was* a chemist – a brilliant one.

Chain and his colleague Howard Florey set about isolating and purifying enough penicillin for further experiments. This required hundreds of litres of mouldy fluid, and their colleague Norman Heatley rigged up a crazy-looking Heath Robinson system involving milk churns, baths, ceramic bedpans commissioned from a local pottery company, rubber tubes, drinks bottles and a doorbell. They employed six women to operate it – the 'penicillin girls'.

The first patient to get an experimental dose was a 43-year-old policeman who'd scratched his cheek while pruning roses and developed septicaemia. Heatley's makeshift system couldn't make penicillin quickly enough, and the police-man died. But, by 1945, penicillin – the first mass-produced antibiotic – was rolling off production lines. Chain, Florey and Fleming shared a Nobel Prize. And Fleming took the opportunity to issue a warning.

'It is not difficult,' Fleming noted, 'to make microbes

resistant to penicillin in the laboratory by exposing them to concentrations not sufficient to kill them.' Fleming worried that an 'ignorant man' might under-dose himself, allowing drug-resistant bacteria to evolve. But ignorance hasn't been the problem. We know the risks, but face incentives to take them anyway.

Suppose I feel ill: perhaps it's viral, meaning antibiotics are useless; even if it's bacterial, I'll probably fight it off. But if there's any chance that antibiotics might speed my recovery, my incentive is to take them. Or suppose I run a pig farm. Giving routine low doses of antibiotics to my pigs is the perfect way to breed antibiotic-resistant bacteria. But that's not my problem. My only incentive is to care about whether dosing my pigs seems to increase my revenues by more than the cost of the drugs. This is a classic example of the tragedy of the commons, where individuals rationally pursuing their own interests ultimately create a collective disaster.

Until the 1970s, scientists kept discovering new antibiotics: when bacteria evolved resistance to one type, we could introduce another. But then the development pipeline dried up. It's possible that new antibiotics will start coming through again: for example, some researchers have come up with a promising new technique to find antimicrobial compounds in soil. Again, though, this is all about incentives. What the world really needs is new antibiotics that we put on the shelf and use only in the direst emergencies. But a product that doesn't get used isn't much of a moneyspinner for drug companies. We'll need to devise better incentives to encourage more research.*

We'll also need smarter regulations as to how new

---

* One possibility: the innovation prizes discussed in Chapter 32 on clocks – and in particular the 'advanced market commitment' funded by five donor governments and by the Gates Foundation. It's already led to the widespread distribution of pneumococcal vaccines.

antibiotics are used, by doctors and farmers alike. Denmark shows it can be done – it's world famous for its bacon, and it strictly controls antibiotic use in pigs. One key appears to be improving *other* regulations, to make farm animals' living conditions less cramped and unhygienic. That makes disease less likely to spread. And recent studies suggest that when animals are kept in better conditions, routine low doses of antibiotics have very little impact on their growth.

The pig farmer in Wuxi meant well. She clearly didn't understand the implications of overusing antibiotics. But even if she had, she'd have faced the same economic incentives to overuse them. Ultimately, that's what needs to change.

42

M-Pesa

When fifty-three policemen in Afghanistan checked their phones, they felt sure there'd been some mistake. They knew they were part of a pilot project, in 2009, to see if public-sector salaries could be paid via a new mobile money service, M-Paisa. But had they somehow overlooked the happy detail that their participation brought a pay rise? Or had someone mistyped the amount of money to send them? The message said their salary was significantly larger than usual.

In fact, the amount was what they should have been getting all along. But previously, they received their salaries in cash, passed down from the ministry via their superior officers. And, somewhere along the line, some of that cash had been getting skimmed off – about 30 per cent. Indeed, the ministry soon realised that one in ten policemen whose salaries they had been dutifully handing over cash for did not, in fact, exist.

The policemen were delighted suddenly to be getting their full salary. Their commanders were less delighted at losing their cut. One was reportedly so irate that he optimistically offered to save his officers the trouble of visiting the M-Paisa agent: just hand over your phones and PINs, and I'll collect your salaries myself.

Afghanistan is among the developing-country economies currently being reshaped by mobile money – the ability to send payments by text message. The ubiquitous kiosks that sell prepaid mobile airtime in effect function like bank branches: you deposit cash, and the agent sends you an SMS adding that amount to your balance, or you send the agent an SMS, and she gives you cash. And you can text some of your balance to anyone else.

It's an invention with roots in many places. But it first took off in Kenya, and that story starts with a presentation made in Johannesburg in 2002. The speaker was Nick Hughes of Vodafone, at the World Summit for Sustainable Development. Nick Hughes's topic was how to encourage large corporations to allocate research funding to ideas that looked risky, but might help poor countries to develop.

In the audience was a man with an answer to that question: an official for DFID, the United Kingdom's Department for International Development. DFID had money to invest in a 'challenge fund' to improve access to financial services. And phones looked interesting: DFID had noticed the customers of African mobile networks were transferring prepaid airtime to each other as a sort of quasi-currency. So the man from DFID had a proposition for Hughes: suppose DFID were to chip in a million pounds, provided Vodafone committed the same, might that help Hughes's ideas to attract the attention of his bosses?

It did. But Hughes's initial idea wasn't about tackling corruption in the public sector, or any of the myriad other imaginative uses to which mobile money is now being put. It was about something much more limited – microfinance, a hot topic in international development at the time. Hundreds of millions of would-be entrepreneurs were too poor for the banking system to bother with, so they couldn't get loans. If only they could borrow a small amount – enough to buy a cow, perhaps, or a sewing machine, or a motorbike – they could start a thriving business. Hughes wanted to explore having microfinance clients repay their loans via SMS.

By 2005, Hughes's colleague Susie Lonie had parked herself in Kenya with Safaricom, a mobile network part-owned by Vodafone. The pilot project didn't always look destined to be a success. Lonie recalls conducting one training session in a sweltering tin shed, battling the noise of a nearby football match and the incomprehension of microfinance clients. Before she could explain M-Pesa, she had to explain how to operate a basic mobile phone. ('Pesa' means 'money' in Kenya, as does 'paisa' in Afghanistan.)

Then people started using the service. And it soon became clear that they were using it for a whole lot more than just repaying loans to microfinance institutions. Intrigued, Lonie dispatched researchers to find out what was going on.

One woman in the pilot project said she'd texted some money to her husband when he was robbed, so he could catch the bus home. Others said they were using M-Pesa to avoid being robbed on the road, depositing money before a journey and withdrawing it on arrival. Businesses were depositing money overnight rather than keeping it in a safe. People were paying each other for services. And workers in the city were using M-Pesa to send money to relatives back home in the village. It was safer than the previous option, entrusting the bus driver with cash in an envelope.

Lonie realised they were onto something big.

Just eight months after M-Pesa launched, a million Kenyans had signed up; that's now about twenty million. Within two years, M-Pesa transfers amounted to 10 per cent of Kenya's GDP – that's since become nearly half. Soon there were a hundred times as many M-Pesa kiosks in Kenya as ATMs.

M-Pesa is a textbook 'leapfrog' technology: where an invention takes hold because the alternatives are poorly developed. Mobile phones allowed Africans to leapfrog their often woefully inadequate landline networks; M-Pesa exposed their banking systems, typically too inefficient to turn a profit from serving the low-income majority. If you're plugged into the financial system, it's easy to take for granted that paying your utility bill doesn't require wasting hours trekking to an office and standing in a queue; or that you have a safer place to accumulate savings than under the mattress. Around two billion people still lack such conveniences, though the number is falling fast – driven largely by mobile money. Most of the poorest Kenyans – those earning under $1.25 a day – had signed up to M-Pesa within a few years.

By 2014, mobile money was in 60 per cent of developing country markets. Some, like Afghanistan, have embraced it quickly – but it hasn't even reached some others. Nor do most developed country customers have the option of sending money by SMS, even though it's simpler than a banking app. Why did M-Pesa take off in Kenya? One big reason was the relaxed approach of the banking and telecoms regulators. Elsewhere, the bureaucrats have not always been as forthcoming.

According to one study, what rural Kenyan households most like about M-Pesa is the convenience for family members sending money home. But two more benefits could be even more profound.

The first was discovered by those Afghan policemen – tackling corruption. In Kenya, similarly, drivers soon realised that the policemen who pulled them over wouldn't take bribes in M-Pesa: it would be linked to their phone number, and could be used as evidence. In many places, corruption is endemic: in Afghanistan, bribes amount to a quarter of GDP. Kenya's matatus, the minibuses that transport people around cities, lose a third of their revenue to theft and extortion.

You might think, then, that matatu operators would have welcomed Kenya's government announcing an ambitious plan to make mobile money mandatory on matatus – after all, if the driver has no cash, he can't be asked for bribes. But many have resisted, and the reason isn't hard to work out. Cash transactions facilitate not only corruption, but also tax evasion. Matatu drivers twigged that when income is traceable, it is also taxable.

That's the other big promise of mobile money: broadening the tax base, by formalising the grey economy. From corrupt police commanders to tax-dodging taxi drivers, mobile money could eventually lead to quite the change in culture.

43

Property Registers

Some of the most important parts of our modern economy are invisible. You can't see radio waves. You can't see limited liability.

And perhaps most fundamentally, you can't see property rights. But you can hear them.

That's what one Peruvian economist concluded about twenty-five years ago, while walking through the idyllic rice fields of Bali, Indonesia. As he passed one farm, a dog would bark at his approach. Then, quite suddenly, the first dog would stop and a new hound would begin to yap away. The boundary between one farm and the next was invisible to him – but the dogs knew exactly where it was. The economist's name is Hernando de Soto. He returned to Indonesia's capital Jakarta, and met with five cabinet ministers to discuss setting up a formal system of property rights. They already knew everything they needed to know, he said, cheekily. They merely needed to ask the dogs of Bali who owned what.

Hernando de Soto is a big name in development economics. His energetic opposition to Peru's Maoist terrorists, The

Shining Path, made him enough of a target that they made three attempts to kill him. And his big idea is to make sure that the legal system can see as much as the dogs of Bali.

But we're getting ahead of ourselves. The Indonesian government was trying to formalise property rights, but many governments have taken the opposite tack. In 1970s China, for example, where the Maoists weren't the rebels but the government, the very idea that anyone could own anything was seditious, bourgeois thinking. Farmers on collective farms were told by Communist Party officials: you don't own a thing. Everything belongs to the collective. What about the teeth in my head, asked one farmer? No, replied the official: even your teeth are owned by the collective.

But this approach worked terribly. If you don't own anything, what incentive is there to work, to invest, to improve your land? Collective farming left farmers in desperate, gnawing poverty. In the village of Xiaogang in 1978, a group of farmers secretly met and agreed to abandon collective farming, divide up the land, and each keep whatever surplus they produced after meeting collective quotas. It was a treasonous agreement in Communist eyes, and the secret contract was hidden away from officials.

But the farmers were eventually found out: the giveaway was that their farms had produced more in one year than in the previous five years combined. It was a tremendously dangerous moment: the farmers were abused and treated like criminals. But, as luck would have it, China had a new leader: Deng Xiaoping. And once Deng let it be known that this was the sort of experiment that had his blessing, 1978 was the beginning of China's breakneck transformation from utter poverty to the largest economy on the planet.

The experience in China shows that property rights are incredibly powerful – and that up to a point, they can be handled informally, by a community. But, says Hernando de Soto, there's a limit to what an informal community agreement can do. If everyone in my neighbourhood recognises that I own my house, that means I can use it in certain important ways. I can sleep there; I can repaint the kitchen – or install a whole new kitchen. If a burglar tries to break in, I can call for help and my neighbours will come.

But in one critical way, it doesn't help me that my neighbours agree that I own my house. It doesn't help me get a loan.

The standard way that anyone raises a serious line of credit is to pledge property as collateral. Land and buildings make particularly good collateral because they tend to increase in value, and because it's hard to hide them from creditors.

But if I want to use my house as collateral for a bank loan, to set up a business or install that new kitchen, I need to prove that the house really is mine. And the bank needs to be confident that it could take the house away from me if I don't repay the loan. To turn a house from a place where I sleep to a place that underpins a business loan requires an invisible web of information that the legal system and the banking system can use.

For Hernando de Soto, this invisible web is the difference between my house being an asset – something useful that I own – and being *capital* – an asset recognised by the financial system.

And it's clear that a lot of assets in poor countries are held informally – de Soto calls them 'dead capital', useless for securing a loan. His estimate was that at the start of the twenty-first century there were almost ten trillion dollars' worth of dead capital across the developing world – more than four thousand dollars for every person. Other researchers think that's an overestimate and the true figure is probably three or four trillion dollars – but it's still clearly a huge amount.

But how do assets become capital? How does the invisible web get woven? Sometimes, it's a top-down affair. Perhaps the first recognisably modern property registry was in Napoleonic France. Napoleon needed to tax things to fund his incessant wars, and property was a good target for taxation. So he decreed that all French properties would be carefully mapped and their ownership would be registered. Such property map is called a 'cadastre', and Napoleon proudly proclaimed that 'a good cadastre of the parcels will be the complement of my civil code'. After conquering Switzerland, the Netherlands and Belgium, Napoleon introduced cadastral maps there, as well.

In the mid-1800s, the idea of the land registry spread quickly through the British Empire, too: state surveyors produced maps, and the department for land would allocate title deeds. It was swift and fairly efficient, and of course at the time nobody with any power had much interest in the fact that most of the allocations were also confiscations from the indigenous people who had their own claims on the land.

In the United States, there was a bottom-up approach. After decades of treating squatters as criminals, the state began to think of them as bold pioneers. The US government tried to formalise informal property claims, using the Pre-emption Act of 1841 and the Homestead Act of 1862. Again, the rights of the native people who had been living there for thousands of years were not regarded as of much significance.

It was hardly justice. But it was profitable. By turning a land-grab into a legally recognised property right, these land registry processes unlocked decades of investment and improvement. And some economists – most prominently

Hernando de Soto himself – argue that the best way to create property registers and cadastral maps for developing countries today is to use the same bottom-up process of recognising informal property rights.

But do improved property registers really unlock what de Soto calls 'dead capital'? The answer, of course, is 'it depends'. It depends on whether there's a banking system capable of lending, and an economy worth borrowing money to invest in.

And it also depends on how smoothly the property register works. De Soto found that in Egypt, legally registering property involved seventy-seven procedures and thirty-one different agencies, and took between five and thirteen years. In the Philippines, everything was twice as complicated: 168 procedures, fifty-three agencies and a thirteen- to twenty-five-year waiting list. Faced with such obstacles, even formally registered properties will soon become informal again – the next time the property is traded, both the buyer and the seller will decide that formalising the deal is just too time consuming.

But get it right, and the results can be impressive. For example, in Ghana, farmers with a clearer right to transfer their property to others invested more in their land. Across the world, the World Bank has found that after controlling for income and economic growth, the countries with simpler, quicker property registries also had less corruption, less grey-market activity, more credit and more private investment.

Property registries occupy a strange place in the political spectrum. On the right, people demand that the government step aside and make space for entrepreneurs. On the left, people urge the government to step forward and involve itself in the economy. Creating and maintaining a property registry is an activity that sits in the overlap of that Venn diagram: if de Soto is right, the government has to act, but it has to act with a minimum of red tape.

Meanwhile, property registries are unfashionable, unloved and even unknown. But without them, many economies would go to the dogs.

VII

INVENTING THE WHEEL

As I wrote in the introduction, this book has been an attempt to identify fifty illuminating stories about how inventions have shaped the modern economy, and decidedly not an attempt to define the fifty most significant inventions in economic history. Nobody could hope to agree on those. But if we tried, one that would surely make most people's lists would be the wheel.

The wheel didn't make the cut in this book, partly because you'd need a book to do it justice. In the modern world we're surrounded by wheels, from the obvious (cars, bikes and trains) to the subtle (the drum in your washing machine, the cooling fan in your computer). Archaeologists believe that the first wheels were probably not – as you might expect – used for transport, but for pottery. One could reasonably credit the wheel for the contents of your crockery cupboard.

But this book does contain plenty of metaphorical wheels: the simple inventions that do a job so well that 'reinventing the wheel' would be foolish. We've seen some of these 'wheels' already – the plough is one; the shipping container is another; so is barbed wire. One of the ultimate 'wheels' is the idea of writing. One can always try to improve on these ideas, but in each case the basic concept has been brilliantly effective.

I have to admit that in writing this book, my favourite inventions have been the wheels.

44

Paper

The Gutenberg printing press: invented in the 1440s by Johannes Gutenberg, a goldsmith from Mainz, it's widely considered to be one of humanity's defining inventions. Gutenberg figured out how to make large quantities of durable metal type, and how to fix that type – firmly enough to print hundreds of copies of a page, yet flexibly enough that the type could then be reused to print an entirely different page. Gutenberg's famous bibles were objects beautiful enough to rival the calligraphy of the monks. You can probably picture a page if you close your eyes – the crisp black Latin script is perfectly composed into two dense blocks of text, occasionally highlighted with a flourish of red ink.

Actually, you can quibble with Gutenberg's place in history. He wasn't the first to invent a movable type press – it was originally developed in China. Even as Gutenberg was inventing his press in the heart of Europe (modern-day Mainz is in western Germany), Koreans were ditching their entire method of writing to make printing easier, cutting tens of thousands of characters down to just twenty-eight. It's often said that Gutenberg single-handedly created mass literacy, but that's not true, either – literacy was common six or seven hundred years earlier in the Abbasid Caliphate, spanning the Middle East and North Africa.

Still, the Gutenberg press changed the world. It led to Europe's reformation, science, the newspaper, the novel, the school textbook and much else. But it could not have done so without another invention, just as essential but much more often overlooked: paper.

Paper was another Chinese idea, from just over two thousand years ago. At first, the Chinese used it for wrapping precious objects. But soon they began to write on it – it was lighter than bamboo, and cheaper than silk. Soon, the Arabic world embraced it.* But Christians in Europe didn't do so until much later; paper came to Germany just a few decades before Gutenberg's press.

What took so long? Europeans, with their danker climate, needed a slightly different kind of paper than that produced by the Arabs. But the real obstacle was lack of demand. For centuries, Europeans just didn't need the stuff. They had parchment, which is made of animal skin. Parchment was pricey: a parchment bible required the skins of 250 sheep. Since so few people could read or write, that hardly mattered. But as a commercial class arose, with more workaday needs like contracts and accounts, the cheaper writing material being used by the Arabs started to look attractive. And the existence of cheap paper made the economics of printing look attractive too: the set-up cost of typesetting could easily be offset with a long print

---

* The word 'ream' – five hundred sheets of paper – is derived from the Arabic 'rizma', meaning bundle or bale. But most English words for paper are dismissive of its origins: both the Latin *papyrus* and the Greek *khartes* (hence cartoon, cartography, card) refer to an Egyptian reed, not to true paper. Papyrus isn't much good for making books, because it frays at creases and edges.

run. That meant either slaughtering a million sheep – or using paper.

And printing is only the start of paper's uses. We decorate our walls with it – as wallpaper or posters and photographs. We filter tea and coffee through it. We package milk and juice in it. As corrugated cardboard, we make boxes with it – or even buildings. An architect named Tina Hovsepian makes 'Cardborigami' buildings – origami-inspired weatherproof cardboard structures that can be folded up, transported, then re-assembled within an hour on the site of a humanitarian emergency.

There's wrapping paper and greaseproof paper and sand-paper. There are paper napkins, paper receipts and paper tickets. And in the 1870s, the same decade that produced the telephone and the light bulb, the British Perforated Paper Company produced a kind of paper that was soft, strong and absorbent – the world's first dedicated toilet paper.

Paper can seem charming and artisanal, but it is the quin-tessential industrial product, churned out at incredible scale. Indeed, when Christian Europeans finally did embrace paper, they created arguably the continent's first heavy industry. Initially, they made paper from pulped cotton. This requires some kind of chemical to break down the raw material – the ammonia from urine works well, so for centuries the paper mills of Europe stank as soiled garments were pulverised in a bath of human piss. The pulping also needs a tremendous amount of mechanical energy: one of the early sites of paper manufacture, Fabriano in Italy, used fast-flowing mountain streams to power massive drop-hammers.

Once finely macerated, the cellulose from the cotton breaks free and floats around in a kind of thick soup; the soup is then thinly poured and allowed to dry, where the cellulose reforms as a strong, flexible mat. Over the years, the process saw innovation after innovation: threshing machines, bleaches, additives – each one designed to make paper more quickly and cheaply, even if the result was often a more fragile surface that yellowed and cracked with age. Paper became an inexpensive product – ideally suited for the everyday jottings of middle-class life. By 1702, paper was so cheap, it was used to make a product explicitly designed to be thrown away after just twenty-four hours: the *Daily Courant*, the world's first daily newspaper.

And then, an almost inevitable industrial crisis: Europe and America became so hungry for paper that they began to run out of rags. The situation became so desperate that scavengers combed battlefields after wars, stripping the dead of their bloodstained uniforms to sell to paper mills. There was an alternative source of cellulose for making paper: wood. The Chinese had long since known how to do it, but the idea was slow to take off in Europe. In 1719, a French biologist, René Antoine Ferchault de Réaumur, wrote a scientific paper pointing out that wasps could make paper nests by chewing up wood, so why couldn't humans? He was ignored for years, and when his idea was rediscovered, papermakers found that wood is not an easy raw material – and it doesn't contain nearly as much cellulose as cotton rags. It was the mid-nineteenth century before wood became a significant source for paper production in the West.

Today, paper is increasingly made out of paper itself – often recycled, appropriately enough, in China. A cardboard box emerges from the paper mills of Ningbo, 130 miles south of Shanghai; it's used to package a laptop computer; the box is shipped across the Pacific; the laptop is extracted, and the box is thrown into a recycling bin in Seattle or Vancouver. Then it's shipped back to Ningbo, to be pulped and turned into another box. The process can be repeated six or seven times before the paper fibres themselves become too short to make strong card.

When it comes to writing, though, some say paper's days are numbered – that the computer will usher in the age of the 'paperless office'. The trouble is, the paperless office has been predicted since Thomas Edison in the late nineteenth century. Remember those wax cylinders, the technology that ushered in recorded music and introduced an age of vast inequality of musicians' incomes? Edison thought they'd be used to replace paper: office memos would be recorded on his wax cylinders instead. Even Edison wasn't right about everything – and when it comes to the death of paper, many other prognosticators have been made to look like fools.

The idea of the paperless office really caught on as computers started to enter the workplace in the 1970s, and it was repeated in breathless futurologists' reports for the next quarter of a century. Meanwhile, paper sales stubbornly continued to boom: yes, computers made it easy to distribute documents without paper, but computer printers made it equally easy for the recipients to put them on paper anyway. America's copiers, fax machines and printers continued to spew out enough sheets of ordinary office paper to cover the country every five years. After a while, the idea of the paperless office became less of a prediction and more of a punchline.

Perhaps things are finally changing: in 2013, the world hit peak paper, with global paper consumption finally starting to decline. Many of us may still prefer the feel of a book or a physical newspaper to swiping a screen, but the cost of digital distribution is now so much lower, we go for the cheaper option. Finally, digital is doing to paper what paper did to parchment with the help of the Gutenberg press: outcompeting it, not on quality, but on price.

Paper may be on the decline, but it's hard to imagine it

disappearing any more than the wheel itself is likely to disappear. It will survive not just on the supermarket shelf or beside the lavatory, but in the office too. Old technologies have a habit of enduring. We still use pencils and candles. The world still produces more bicycles than cars. Paper was never just a home for the beautiful typesetting of a Gutenberg Bible; it was everyday stuff. And for jottings, lists and doodles, you still can't beat the back of an envelope.

45

Index Funds

Here's a question: what's the best financial investment in the world?

If anyone knows the answer, it's Warren Buffett – the world's richest investor and one of the world's richest people, full stop. He's worth tens of billions of dollars, accumulated over decades of savvy investments. And Warren Buffett's advice? It's in a letter he wrote to his wife, advising her how to invest after he dies. And it's been published online for anyone to read.

Those instructions: pick the most mediocre investment you can imagine. Put almost everything into 'a very low-cost S&P 500 index fund'.

Yes. An index fund. The idea of an index fund is to be mediocre by definition – to passively track the market as a whole by buying a little of everything, rather than trying to beat the market with clever stock picks – the kind of clever stock picks that Buffett himself has been making for more than half a century.

Index funds now seem completely natural – part of the very language of investing. But as recently as 1976, they didn't exist.

Before you can have an index fund, you need an index. In 1884, a financial journalist called Charles Dow had the bright idea that he could take the price of some famous company stocks and average them, then publish the average going up and down. He ended up founding not only the Dow Jones company, but also the *Wall Street Journal*.

The Dow Jones Industrial Average didn't pretend to do anything much except track how shares were doing, as a whole. But thanks to Charles Dow, pundits could now talk about the stock market rising by 2.3 per cent or falling by 114 points. More sophisticated indices followed – the Nikkei, the Hang Seng, the Nasdaq, the FTSE and, most famously, the S&P 500. They quickly became the meat and drink of business reporting all around the world.

Then, in the autumn of 1974, the world's most famous economist took an interest. The economist's name was Paul Samuelson. He had revolutionised the way economics was practised and taught, making it more mathematical – more like engineering and less like a debating club. His book, *Economics*, was America's bestselling textbook in any subject for almost thirty years. He advised President Kennedy. He won one of the first Nobel memorial prizes in economics.

Samuelson had already proved the most important idea in financial economics: that if investors were thinking rationally about the future, the price of assets like shares and bonds should fluctuate randomly. That seems paradoxical, but the intuition is that all the predictable movements have already happened: lots of people will buy a share that's obviously a bargain, and then the price will rise and it won't be an obvious bargain any more.

That idea has become known as the efficient markets hypothesis. It's probably not quite true – investors aren't perfectly rational and some investors are more interested in covering their backsides than taking well-judged risks. But the efficient markets hypothesis is true-ish. And the truer it is, the harder it's going to be for anyone to beat the stock market.

Samuelson looked at the data, and found – embarrassingly for the investment industry – that indeed, in the long run most professional investors didn't beat the market. And while some did, good performance often didn't last. There's a lot of luck involved and it's hard to distinguish that luck from skill.

Samuelson laid out his thinking in an article called 'Challenge to Judgement', which said that most professional investors should quit and do something useful instead, like plumbing. And Samuelson went further: he said that since professional investors didn't seem to be able to beat the market, somebody should set up an index fund – a way for ordinary people to invest in the performance of the stock market as a whole, without paying a fortune in fees for fancy professional fund managers to try, and fail, to be clever.

At this point, something interesting happened: a practical businessman actually paid attention to what an academic economist had written. John Bogle had just founded a company called Vanguard, whose mission was to provide simple mutual funds for ordinary investors – no fuss, no fancy stuff, low fees. And what could be simpler and cheaper than an index fund – as recommended by the world's most respected economist? And so Bogle decided he was going to make Paul Samuelson's wish come true. He set up the world's first index fund, and waited for investors to rush in.

They didn't. When Bogle launched the First Index Investment Trust in August 1976, it flopped. Investors weren't interested in a fund that was guaranteed to be mediocre. Financial professionals hated the idea – some even said it was un-American. It was certainly a slap in their faces: Bogle was in effect saying, 'don't pay these guys to pick stocks, because they can't do better than random chance. Neither can I, but at least I charge less.' People called Vanguard's index fund 'Bogle's Folly'.

But Bogle kept the faith, and slowly people started to catch on. Active funds are expensive, after all. They often trade a lot, buying and selling stocks in search of bargains. They pay analysts handsomely to fly around meeting company directors. The annual fees might sound modest – just 1 or 2 per cent – but they soon mount up: if you're saving for retirement, a 1 per cent annual fee could easily eat away a quarter or more of your retirement fund. Now if the analysts consistently outperform the market, then even that sort of fee would be money well spent. But Samuelson showed that, in the long run, most don't.

The super-cheap index funds looked, over time, to be a perfectly credible alternative to active funds – without all the costs. So, slowly and surely, Bogle's funds grew and spawned more and more imitators – each one passively tracking some broad financial benchmark or other, each one tapping into Paul Samuelson's basic insight that if the market is working well you might as well sit back and go with the flow. Forty years after Bogle launched his index fund, fully 40 per cent of US stock market funds are passive trackers rather than active stock-pickers. You might say that the remaining 60 per cent are clinging to hope over experience.

Index investing is a symbol of the power of economists to change the world that they study. When Samuelson and his successors developed the idea of the efficient markets hypothesis, they changed the way that markets themselves worked – for better or worse. It wasn't just the index fund. Other financial products such as derivatives really took off after economists figured out how to value them. Some scholars think the efficient markets hypothesis itself played a part in the financial crisis, by encouraging something called 'mark to market' accounting – where a bank's accountants would figure out what its assets were worth by looking at their value on financial markets. There's a risk that such accounting leads to self-reinforcing booms and busts, as everyone's books suddenly and simultaneously look brilliant, or terrible, because financial markets have moved.

Samuelson himself, understandably, thought that the index fund had changed the world for the better. It's already saved ordinary investors literally hundreds of billions of dollars. That's a big deal: for many, it'll be the difference between scrimping and saving or relative comfort in old age. In a speech in 2005, when Samuelson himself was ninety years old, he gave Bogle the credit. He said: 'I rank this Bogle invention along with the invention of the wheel, the alphabet, Gutenberg printing, and wine and cheese: a mutual fund that never made Bogle rich but elevated the long-term returns of the mutual-fund owners. Something new under the sun.'

46

The S-bend

'Gentility of speech is at an end,' thundered an editorial in London's *City Press* in 1858. 'It stinks.'

The stink in question was partly metaphorical: politicians were failing to tackle an obvious problem. As London's population grew, the city's system for disposing of human waste became woefully inadequate. To relieve pressure on cesspits – which were prone to leaking, overflowing and belching explosive methane – the authorities had instead started encouraging sewage into gullies. However, this created a different issue: the gullies were originally intended only for rainwater, and they emptied directly into the River Thames.

That was the literal stink – the Thames became an open sewer. The distinguished scientist Michael Faraday was moved by a boat journey to write to *The Times*. He described the river water as 'an opaque, pale brown fluid ... [n]ear the bridges the feculence rolled up in clouds so dense that they were visible at the surface.' The smell, he said, was 'very bad ... the same as that which now comes up from the gulley holes in the streets.'

Cholera was rife. One outbreak killed fourteen thousand Londoners – nearly one in every hundred. Civil engineer Joseph Bazalgette drew up plans for new, closed sewers to pump the waste far from the city. It was this project that politicians came under pressure to approve.

Faraday ended his letter by pleading with 'those who exercise power or have responsibility' to stop neglecting the problem, lest 'a hot season gives us sad proof of the folly of our carelessness'. And, three years later, that's exactly what happened. The sweltering-hot summer of 1858 made London's malodorous river impossible to politely ignore or to discuss obliquely with 'gentility of speech'. The heatwave became popularly known as the 'Great Stink'.

If you live in a city with modern sanitation, it's hard to imagine daily life being permeated with the suffocating stench of human excrement. For that, we have a number of people to thank – but perhaps none more so than the unlikely figure of Alexander Cumming. A watchmaker in London a century before the Great Stink, Cumming was renowned for his mastery of intricate mechanics: he served as a judge for the Longitude prize that spurred John Harrison's development of the world's best timekeeping device. King George III commissioned Cumming to make an elaborate instrument for recording atmospheric pressure, and he pioneered the microtome, a device for cutting ultra-fine slivers of wood for microscopic analysis.

But Cumming's world-changing invention owed nothing to precision engineering. It was a bit of pipe with a curve in it.

In 1775, Cumming patented the S-bend. This became the missing ingredient to create the flushing toilet – and, with it, public sanitation as we know it. Flushing toilets had previously foundered on the problem of smell: the pipe that connects the toilet to the sewer, allowing urine and faeces to be flushed away, will also let sewer odours waft back up – unless you can create some kind of airtight seal.

Cumming's solution was simplicity itself: bend the pipe. Water settles in the dip, stopping smells coming up; flushing the toilet replenishes the water. While we've moved on alphabetically from the S-bend to the U-bend, flushing toilets still deploy the same insight: Cumming's invention was almost unimprovable.

Roll-out, however, came slowly: by 1851, flushing toilets remained novel enough in London to cause mass excitement when they were introduced at the Great Exhibition in Crystal Palace. Use of the facilities cost one penny, giving the English language one of its enduring euphemisms for emptying one's bladder: 'to spend a penny'. Hundreds of thousands of Londoners queued for the opportunity to relieve themselves while marvelling at the miracles of modern plumbing.

If the Great Exhibition gave Londoners a vision of how public sanitation could be – clean, and odour-free – no doubt that added to the weight of popular discontent as politicians dragged their heels over finding the funds for Bazalgette's planned sewers. Those plans weren't perfect. At the time, it was wrongly believed that smell caused disease, so Bazalgette assumed it would suffice to pump the sewage into the Thames further downstream. As it happened, that did largely solve the actual cause of cholera – contaminated drinking water – but it didn't help if you wanted to fish in the estuary or bathe on a nearby beach. Cities that are currently experiencing infrastructure-stretching population explosions now have much more knowledge to draw on than was available in 1850s London.

But we still haven't reliably managed to solve the problem of collective action, how to get 'those who exercise power or have responsibility', as Faraday put it, to organise themselves.

There has been a great deal of progress. According to the World Health Organisation, the proportion of the world's people who have access to what's called 'improved sanitation' has increased from around a quarter in 1980 to around two thirds today. That's a big step forward.

Still, 2.5 billion people remain without improved sanitation, and improved sanitation itself is a low bar: the definition is that it 'hygienically separates human excreta from human contact' but it doesn't necessarily treat the sewage itself. Fewer than half the world's people have access to sanitation systems that do that.

The economic costs of this ongoing failure to roll out proper sanitation are many and varied, from healthcare for diarrhoeal diseases to forgone revenue from hygiene-conscious tourists. The World Bank's 'Economics of Sanitation Initiative' has tried to tot up the price tag – across various African countries, for example, it reckons inadequate sanitation lops 1 or 2 percentage points off GDP; in India and Bangladesh, over 6 per cent; in Cambodia, 7 per cent. That soon adds up: countries that *have* made good use of Cumming's S-bend are now a whole lot richer for it.

The challenge is that public sanitation isn't something the market necessarily provides. Toilets cost money, but defecating in the street is free. If I install a toilet, I bear all the costs, while the benefits of the cleaner street are felt by everyone. In economic parlance, that's what is known as a positive externality – and goods that have positive externalities tend to be bought at a slower pace than society, as a whole, would prefer.

The most striking example is the 'flying toilet' system of Kibera, a famous slum in Nairobi, Kenya. The flying toilet works like this: you poo into a plastic bag, and then in the middle of the night, whirl the bag around your head and hurl it as far away as possible. Replacing a flying toilet with a flushing toilet provides benefits to the toilet owner – but you can bet that the neighbours would appreciate it, too.

Contrast, say, the mobile phone. That doesn't have as many positive externalities. It does have some: if I buy a phone, my neighbours with phones can contact me more easily, which benefits them. Given the choice, though, they'd surely prefer that I instead refrain from throwing excrement around. But most of the benefits of owning a phone accrue to me. Suppose, then, that I'm deciding whether to buy a phone or save that money towards a toilet: if I altruistically add up the benefits to myself and my neighbours, I might decide on the toilet; if I selfishly look only at the benefits to myself, I might prefer the phone. That's one reason why, although the S-bend has been around for ten times as long as the mobile phone, many more people currently own a mobile phone than a flushing toilet.

In Kibera, efforts to eliminate the flying toilets have centred around a small number of communal washroom blocks – and the distribution of specially designed toilet bags that can be filled, collected and used for compost. It's an ad hoc solution to the constraints of Kibera.

Modern sanitation requires more than a flushing toilet, of course. It helps if there's a system of sewers to plumb it into, and creating one is a major undertaking – financially and logistically. When Bazalgette finally got the cash to build London's sewers, they took ten years to complete and necessitated digging up 2.5 million cubic metres of earth. Because of the externality problem, such a project might not appeal to private investors: it tends to require determined politicians, willing taxpayers and well-functioning municipal governments to accomplish. And those are in short supply. India, for example, has 5161 towns or cities, according to its latest census. How many have succeeded in constructing even a partial network of sewers? Less than 6 per cent.

London's lawmakers likewise procrastinated – but when they finally acted, they didn't hang about. It took just eighteen days to rush through the necessary legislation for Bazalgette's plans. As we've seen, whether it's deregulating the US trucking industry, reforming property registers in Peru, or making sure that banking doesn't destabilise the economy, it's not easy to get politicians to act with speed and wisdom. So what explains this remarkable alacrity?

A quirk of geography: London's parliament is located next to the River Thames. Officials tried to shield lawmakers from the Great Stink, soaking the building's curtains in chloride of lime in a bid to mask the stench. But it was no use – try as they might, the politicians couldn't ignore it. *The Times* described, with a note of grim satisfaction, how Members of Parliament had been seen abandoning the building's library, 'each gentleman with a handkerchief to his nose'. If only concentrating politicians' minds were always that easy.

47

Paper Money

Almost seven hundred and fifty years ago, a young Venetian merchant named Marco Polo wrote a remarkable book chronicling his travels in China. It was called *The Book of the Marvels of the World*, and it was full of strange foreign customs that Marco claimed to have seen. But there was one, in particular, that was so extraordinary, he could barely contain himself: 'tell it how I might,' he wrote, 'you never would be satisfied that I was keeping within truth and reason.'

What was exciting Marco so much? He was one of the first Europeans to witness an invention that remains at the foundation of the modern economy: paper money.

Of course, the paper itself isn't the point. Modern paper money isn't made of paper – it's made of cotton fibres or a flexible plastic weave. And the Chinese money that so fascinated Marco Polo wasn't quite paper either. It was made from a black sheet derived from the bark of mulberry trees, signed by multiple officials and, with a seal smothered in bright red vermilion, authenticated by the Chinese emperor Genghis Khan himself. The chapter of Marco Polo's book was titled, somewhat breathlessly: 'How the great Khan Causes the Bark of Trees, Made into Something Like Paper, to Pass for Money All over His Country'.

The point is that whatever these notes were made of, their value didn't come from the preciousness of the substance, as with a gold or silver coin. Instead, the value was created purely by the authority of the government. Paper money is sometimes called fiat money – the Latin word *fiat* means 'let it be done'. The Great Khan announces that officially stamped mulberry bark is money – and lo, let it be done. Money it is.

The genius of this system amazed Marco Polo, who explained that the paper money circulated as though it was gold or silver itself. Where was all the gold that wasn't circulating? Well, the emperor kept a tight hold of that.

The mulberry money itself wasn't new when Marco Polo heard about it. It had emerged nearly three centuries earlier, around the year AD 1000 in Sichuan, China – a region now best known for its fiery cuisine. Sichuan was a frontier province, bordered by foreign and sometimes hostile states. China's rulers didn't want valuable gold and silver coins to leak out of Sichuan into foreign lands, and so they told Sichuan to use coins made of iron.

Iron coins aren't terribly practical. If you traded in a handful of silver coins – just 50 grams' worth – you'd be given your own body weight in iron coins. Even something simple like salt was worth more, gram for gram, than iron – so if you went to the market for groceries, your sackful of coins on the way there would weigh more than the bag of goods that you brought back.

It's no surprise that they began to experiment with an alternative.

That alternative was called *jiaozi*, or 'exchange bills'. They were simply IOUs. Instead of carrying around a wagonload of iron coins, a well-known and trusted merchant would write an IOU, and promise to pay his bill later when it was more convenient for everyone.

But then something unexpected happened.

These IOUs, these *jiaozi*, started to trade freely. Suppose I supply some goods to the eminently reputable Mr Zhang, and he gives me an IOU. When I go to your shop later, rather than paying you with iron coins – who does that? – I could write you an IOU. But it might be simpler – and indeed you might prefer it – if instead I give you Mr Zhang's IOU. After all, we both know that Mr Zhang is good for the money.

Now you, and I, and Mr Zhang, have together created a kind of primitive paper money – it's a promise to repay that has a marketable value of its own and can be passed around from person to person without being redeemed. The very idea of this is slightly bewildering at first, but as we saw in Chapter 20, this sort of tradable debt has emerged at other times – the cheques passed around Ireland during the banking strike of the 1970s, or in 1950s Hong Kong, or even the willow tally sticks of late medieval England.

The new system of tradable promises is very good news for Mr Zhang, because as long as one person after another finds it convenient simply to pass on his IOU as a way of paying for things, Mr Zhang never actually has to stump up the iron coins. In effect, he enjoys an interest-free loan for as long as his IOU continues to circulate. Better still, it's a loan that he may never be asked to repay.

No wonder the Chinese authorities started to think these benefits ought to accrue to them, rather than to the likes of Mr Zhang. At first they regulated the issuance of *jiaozi* and produced rules about how they should look. Soon enough, the authorities outlawed private *jiaozi* and took over the whole business themselves. The official *jiaozi* currency was a huge hit; it circulated across regions and even internationally.

In fact, the *jiaozi* even traded at a premium, because they were so much easier to carry around than metal coins.

Initially, government-issued *jiaozi* could be redeemed for coins on demand, exactly as the private *jiaozi* had been. This is a logical enough system: it treats the paper notes as a placeholder for something of real value. But the government soon moved stealthily to a fiat system, maintaining the principle but abandoning the practice of redeeming *jiaozi* for metal. Bring an old *jiaozi* in to the government treasury to be redeemed, and you would receive . . . a crisp new *jiaozi*.

That was a very modern step. The money we use today all over the world is created by central banks and it's backed by nothing in particular except the promise to replace old notes with fresh ones. We've moved from a situation where Mr Zhang's IOU circulates without ever being redeemed, to the mind-bending situation where the government's IOUs circulate despite the fact they *cannot* be redeemed.

For governments, fiat money represents a temptation: a government with bills to pay can simply print more money. And when more money chases the same amount of goods and services, prices go up. The temptation quickly proved too great to resist. Within a few decades of its invention in the early eleventh century, the *jiaozi* was devalued and discredited, trading at just 10 per cent of its face value.

Other countries have since suffered much worse. Weimar Germany and Zimbabwe are famous examples of economies collapsing into chaos as excessive money-printing rendered prices meaningless. In Hungary in 1946, prices trebled every day. Walk into a Budapest café back then, and it was better to pay for your coffee when you arrived, not when you left.

These rare but terrifying episodes have convinced some economic radicals that fiat money can never be stable: they yearn for a return to the days of the gold standard, when paper money could be redeemed for a little piece of precious metal. But mainstream economists generally now believe that pegging the money supply to gold is a terrible idea. Most regard low and predictable inflation as no problem at all – perhaps even as a useful lubricant to economic activity because it guards against the possibility of deflation, which can be economically disastrous. And while we may not always be able to trust central bankers to print just the right amount of new money, it probably makes more sense than trusting miners to dig up just the right amount of new gold.

The ability to fire up the printing presses is especially useful in crisis situations. After the 2007–8 financial crisis, the US Federal Reserve pumped trillions of dollars into the economy, without creating inflation. In fact, the printing presses were metaphorical: those trillions were created by key-strokes on computers in the global banking system. As a wide-eyed Marco Polo might have put it: 'The great Central Bank Causes the Digits on a Computer Screen, Made into Something Like Spreadsheets, to Pass for Money'. Technology has changed, but what passes for money continues to astonish.

48

Concrete

In the early part of the twenty-first century, poor families in the Coahuila state in Mexico were offered an unusual handout from a social programme called *Piso Firme*. It wasn't a place at school, a vaccination, food, or even money. It was 150 dollars' worth of ready-mixed concrete. Workers would drive concrete mixers through poor neighbourhoods, stop outside the home of a needy family, and pour out the porridge-like mixture, through the door, right into the living room. Then they would show the occupants how to spread and smooth the gloop, and make sure they knew how long to leave it to dry. And they'd drive off to the next house.

Piso Firme means 'firm floor', and when economists studied the programme they found that the ready-mixed concrete dramatically improved children's education. How so? Previously, most house floors were made of dirt. Parasitic worms thrive in dirt, spreading diseases that stunt kids' growth and make them sick. Concrete floors are much easier to keep clean: so the kids were healthier, they went to school more regularly, and their test scores improved. Living on a dirt floor is unpleasant in many other ways: economists also found that parents in the programme's households became happier, less stressed and less prone to depression when they lived in houses with concrete floors. That seems to be 150 dollars well spent.

Beyond the poor neighbourhoods of Coahuila state, concrete often has a less wonderful reputation. It's a byword for ecological carelessness: concrete is made of sand, water and cement, and cement takes a lot of energy to produce; the production process also releases carbon dioxide, a greenhouse gas. That might not be such a problem in itself – after all, steel production needs a lot more energy – except that the world consumes absolutely vast quantities of concrete: 5 tonnes, per person, per year. As a result the cement industry emits as much greenhouse gas as aviation. Architecturally, concrete implies lazy, soulless structures: ugly office blocks for provincial bureaucrats, or multi-storey car parks with stairwells that smell of urine. Yet it can also be shaped into forms that many people find beautiful – think of the Sydney Opera House, or Oscar Niemeyer's cathedral in Brasilia.

Perhaps it's no surprise that concrete can evoke such confusing emotions. The very nature of the stuff feels hard to pin down. 'Is it Stone? Yes and No,' opined the great American architect Frank Lloyd Wright in 1927. He continued, 'Is it Plaster? Yes and No. Is it Brick or Tile? Yes and No. Is it Cast Iron? Yes and No.'

That it's a great building material, however, has been recognised for millennia – perhaps even since the dawn of human civilisation. There's a theory that the very first settlements, the first time that humans gathered together outside their kinship groups – nearly twelve thousand years ago at Göbekli Tepe in southern Turkey – was because someone had figured out how to make cement, and therefore concrete. It was certainly being used over eight thousand years ago by desert traders to make secret underground cisterns to store scarce water; some of these cisterns still exist in modern-day Jordan and Syria. The Mycenaeans used it over three thousand years ago, to make tombs you can see in the Peloponnese in Greece.

The Romans were serious about the stuff. Using a naturally occurring cement from volcanic ash deposits at Puteoli, near Pompeii and Mount Vesuvius, they built their aqueducts and their bathhouses with concrete. Walk into the Pantheon in Rome, a building that will soon celebrate its nineteen-hundredth birthday. Gaze up at what was the largest dome on the planet for centuries, arguably until 1881. You're looking at concrete. It's shockingly modern.

Many Roman brick buildings are long gone – but not because the bricks themselves have decayed. They've been cannibalised for parts. Roman bricks can be used to make modern buildings. But the concrete Pantheon? One of the reasons it has survived for so long is because the solid concrete structure is absolutely useless for any other purpose. Bricks can be reused; concrete can't. It can only be reduced to rubble. And the chances of it becoming rubble depend on how well it's made. Bad concrete – too much sand, too little cement – is a deathtrap in an earthquake. But well-made concrete is waterproof, stormproof, fireproof, strong and cheap.

That's the fundamental contradiction of concrete: incredibly flexible while you're making something, utterly inflexible once it's made. In the hands of an architect or a structural engineer, concrete is a remarkable material – you can pour it into a mould, and set it to be slim and stiff and strong in almost any shape you like. It can be dyed, or grey; it can be rough or polished smooth like marble. But the moment the building is finished, the flexibility ends: cured concrete is a stubborn, unyielding material.

Perhaps that is why the material has become so associated with arrogant architects and autocratic clients – people who believe that their visions are eternal, rather than likely to need deconstructing and reconstructing as times and circumstances change. In 1954 the then Soviet leader Nikita Khrushchev delivered a two-hour speech praising concrete, and proposing in some detail his ideas for standardising it still further. He wanted to embrace 'a single system of construction for the whole country'. No wonder we think of concrete as something that is imposed upon people, not something they choose for themselves.

Concrete is permanent and yet throwaway. It lasts forever. In a million years, when our steel has rusted and our wood has rotted, concrete will remain. But many of the concrete structures we're building today will be useless within decades. That's because, over a century ago, there was a revolutionary improvement in concrete – but it's an improvement with a fatal flaw.

In the mid-nineteenth century, a French gardener, Joseph Monier, was unsatisfied with the available range of flower pots. Concrete pots had become fashionable, but they were either brittle or bulky. Customers loved the modern look, but Monier didn't want to have to lug around cumbersome planters, and so he experimented with pouring concrete over a steel mesh. It worked brilliantly.

Monier had been rather lucky. Reinforcing concrete with steel simply shouldn't work, because different materials tend to expand by different amounts when they warm up. A concrete flower pot in the sun should crack, as the concrete expands by a certain amount and the steel inside expands at a different rate. But by a splendid coincidence, concrete and steel expand in a similar way when they heat up – they're a perfect pairing.

But Monier rode his luck: over time, he realised that reinforced concrete had many more applications besides flower pots – railway sleepers, building slabs and pipes – and he patented several variants of the invention, which he exhibited at the Paris International Exhibition in 1867.

Other inventors took up the idea, testing the limits of reinforced concrete and working out how to improve it. Less than twenty years after Monier's first patent, the elegant idea of pre-stressing the steel was patented. The pre-stressing makes the concrete stronger – it partly counteracts the forces that will act on the concrete when it's in use. Pre-stressing allows engineers to use much less steel, and less concrete too. And it works as well as ever a hundred and thirty years later.

Reinforced concrete is much stronger and more practical than the unreinforced stuff. It can span larger gaps, allowing concrete to soar in the form of bridges and skyscrapers. But here's the problem: if it is made cheaply, it will rot from the inside as water gradually seeps in through tiny cracks, and rusts the steel. This process is currently destroying infrastructure across the United States;* in twenty or thirty years' time, China will be next. China poured more concrete in the three years after 2008 than the United States poured during the entire twentieth century, and nobody thinks that all that concrete is made to exacting standards.

There are many schemes to improve concrete, including special treatments to prevent water getting through to the steel. There's 'self-healing' concrete, full of bacteria that secrete limestone, which re-seals any cracks. And 'self-cleaning'

---

* The American Society for Civil Engineers noted in its '2013 Report Card for America's Infrastructure' that 'one in nine of the nation's bridges are rated as structurally deficient' and 'The Federal Highway Administration (FHWA) estimates that to eliminate the nation's bridge backlog by 2028, we would need to invest $20.5 billion annually, while only $12.8 billion is being spent currently.'

concrete, infused with titanium dioxide, which breaks down smog, keeping the concrete sparkling white; improved versions of the technology may even give us street surfaces that clean what's coming out of car exhausts.

Scientists are figuring out ways to reduce energy use and carbon emissions in making concrete. If they succeed in that, the environmental rewards will be high.

Yet ultimately, there is much more we could be doing with the simple, trusted technology we have already. Hundreds of millions of people around the world live in dirt-floor houses; hundreds of millions of people could have their lives improved by a programme like *Piso Firme*. Other studies have shown large gains from laying concrete roads in rural Bangladesh – improving school attendance, agricultural productivity and the wages of farm workers.

Perhaps concrete serves us best when we use it simply.

49

Insurance

A decade ago – as a stunt for a radio programme – I phoned up one of the UK's leading betting shops and tried to take a bet with them that I was going to die. They refused, which cost them money since I am, after all, still alive. But a betting shop won't gamble on life and death. A life insurance company, by contrast, does little else.

Legally and culturally, there's a clear distinction between gambling and insurance. Economically the difference is not so easy to see. Both the gambler and the insurer are agreeing that money will change hands depending on what transpires in some unknowable future.

It's an old – almost primal – idea. Gambling tools such as dice date back millennia; perhaps five thousand years in Egypt. In India, twenty-five centuries ago, they were popular enough to be included in a list of games that the Buddha refused to play. Insurance may be equally ancient. The Code of Hammurabi – a law code from Babylon, in what is now Iraq – is nearly four thousand years old. It devotes much attention to the topic of 'bottomry', which was a kind of maritime insurance bundled together with a business loan: a merchant would borrow money to fund a ship's voyage, but if the ship sank, the loan did not have to be repaid.

Around the same time, Chinese merchants were spreading their risks by swapping goods between ships – if any one ship went down, it would contain a mix of goods from many different merchants. But all that physical shuffling around is a fuss. It's more efficient to structure insurance as a financial contract instead. A couple of millennia later, that's what the Romans did, with an active marine insurance market. Later still, Italian city states such as Genoa and Venice continued the practice, developing increasingly sophisticated ways to insure the ships of the Mediterranean.

Then, in 1687, a coffee house opened on Tower Street, near the London docks. It was comfortable and spacious, and business boomed. Patrons enjoyed the fire; tea, coffee and sherbet; and of course, the gossip. There was a lot to gossip about: London had recently experienced the plague, the great fire, the Dutch navy sailing up the Thames and a revolution that had overthrown the king.

But above all, the inhabitants of this coffee house loved to gossip about ships: what was sailing from where, with what cargo – and whether it would arrive safely or not. And where there was gossip, there was an opportunity for a wager. The patrons loved to bet. They bet, for example, on whether Admiral John Byng would be shot for his incompetence in a naval battle with the French; he was. The patrons of this coffee house would have had no qualms about taking my bet on my own life.

The proprietor saw that his customers were as thirsty for information to fuel their bets and gossip as they were for coffee, and so he began to assemble a network of informants and a newsletter full of information about foreign ports, tides and the comings and goings of ships. His name was Edward

Lloyd. His newsletter became known as Lloyd's List. Lloyd's coffee house hosted ship auctions and gatherings of sea captains who would share stories. And if someone wished to insure a ship, that could be done too: a contract would be drawn up, and the insurer would sign his name underneath – hence the term 'underwriter'. It became hard to say quite where coffee-house gambling ended and formal insurance began.

Naturally, underwriters congregated where they could be best informed, because they needed the finest possible grasp of the risks they were buying and selling. Eight decades after Lloyd had established his coffee house, a group of underwriters who hung out there formed the Society of Lloyd's. Today, Lloyd's of London is one of the most famous names in insurance.

But not all modern insurers have their roots in gambling. Another form of insurance developed not in the ports but in the mountains – and rather than casino capitalism, this was community capitalism. Alpine farmers organised mutual aid societies in the early sixteenth century, agreeing to look after each other if a cow – or perhaps a child – got sick. While the underwriters of Lloyd's viewed risk as something to be analysed and traded, the mutual assurance societies of the Alps viewed risk as something to be shared. A touchy-feely view of insurance, perhaps, but when the farmers descended from the Alps to Zurich and Munich, they established some of the world's great insurance companies.

Risk-sharing mutual aid societies are now among the largest and best-funded organisations on the planet: we call them 'governments'. Governments initially got into the insurance business as a way of making money, typically to fight some war or other in the turmoil of Europe in the 1600s and 1700s. Instead of selling ordinary bonds, which paid in regular instalments until they expired, governments sold annuities, which paid in regular instalments until the *recipient* expired. Such a product was easy enough for a government to supply, and they were much in demand. Annuities were popular products because they, too, are a form of insurance – they insure an individual against the risk of living so long that all her money runs out.

Providing insurance is no longer a mere money-spinner for governments. It is regarded as one of their core priorities to help citizens manage some of life's biggest risks – unemployment, illness, disability and ageing. Much of the welfare state, which we discussed back in Chapter 8, is really just a form of insurance. In principle some of this insurance could be provided by the marketplace, but faced with such deep pools of risk, private insurers often merely paddle. And in poorer countries, governments aren't much help against life-altering risks, such as crop failure or illness. Private insurers don't take much of an interest, either. The stakes are too low, and the costs too high.

That is a shame. The evidence is growing that insurance doesn't just provide peace of mind, but is a vital element of a healthy economy. For example, a recent study in Ghana showed that highly productive farmers were being held back from specialising and expanding by the risk of drought – a risk against which they couldn't insure themselves. When researchers created an insurance company and started selling crop insurance, the farmers bought the product and expanded their businesses.

But for private insurers, there isn't much money to be made supplying crop insurance in Ghana. They can do better by playing on our fears of the slings and arrows of outrageous fortune and selling richer consumers overpriced cover against overblown risks, like a cracked mobile phone screen.

Today, the biggest insurance market of all blurs the line between insuring and gambling: the market in financial derivatives. Derivatives are financial contracts that let two parties bet on something else – perhaps exchange rate fluctuations, or whether a debt will be repaid or not. They can be a form of insurance. An exporter hedges against a rise in the exchange rate; a wheat farming company covers itself by betting that the price of wheat will fall. For these companies, the ability to buy derivatives frees them up to specialise in a particular market. Otherwise, they'd have to diversify – like the Chinese merchants four millennia ago, who didn't want all their goods in one ship. And the more an economy specialises, the more it tends to produce.

But unlike boring old regular insurance, for derivatives you don't need to find someone with a risk they need to protect themselves against: you just need to find someone willing to take a gamble on any uncertain event anywhere in the world. It's a simple matter to double the stakes – or multiply them by a hundred. As the profits multiply, all that's needed is the appetite to take risks. In 2007, before the international banking crisis broke, the total face value of outstanding derivatives contracts was many times larger than the world economy itself. The real economy became the sideshow; the side bets became the main event. That story did not end well.

CONCLUSION

Looking Forward

E conomic calamities like the global financial crisis of 2007–8 should not obscure the bigger picture: life for most people today is vastly better than it was for most people in the past.

A century ago, for example, global average life expectancy at birth was just thirty-five; when I was born, it was sixty; recently it rose above seventy. A baby born in the least propitious countries today such as Burma, Haiti and the Democratic Republic of Congo has a better chance of surviving infancy than any baby born in 1900. The proportion of the world's population living in the most extreme poverty has fallen from about 95 per cent two centuries ago to about 60 per cent fifty years ago to about 10 per cent today.

Ultimately, the credit for this progress rests with the invention of new ideas like the ones described in this book. And yet, few of the stories we've had to tell about these inventions have been wholly positive. Some inventions did great harm. Others would have done much more good if we had used them wisely.

It's reasonable to assume that future inventions will deliver a similar pattern: broadly, they will solve problems and make us richer and healthier, but the gains will be uneven and there will be blunders and missed opportunities.

It's fun to speculate about what those inventions might be, but history cautions against placing much faith in futurology. Fifty years ago, Herman Kahn and Anthony J. Wiener published *The Year 2000: A Framework For Speculation*. Their crystal-ball gazing got a lot right about information and communication technology. They predicted colour photocopying, multiple uses for lasers, 'two-way pocket phones' and automated real-time banking. That's impressive. But Kahn and Wiener also predicted undersea colonies, silent helicopter-taxis and cities lit by artificial moons. Nothing looks more dated than yesterday's technology shows and yesterday's science fiction.

We can make two predictions, though. First, the more human inventiveness we encourage, the better that's likely to work out for us. And, second, with any new invention, it makes sense to at least ask ourselves how we might maximise the benefits and mitigate the risks.

What lessons have we learned from the forty-nine inventions so far?

We're already a long way towards learning one big lesson about encouraging inventiveness: most societies have realised that it isn't sensible to waste the talents of half their populations. It will not have escaped your notice that most of the inventors we've encountered have been male, and no wonder – who knows how many brilliant women, like Clara Immerwahr, were lost to history after having their ambitions casually crushed.

Education matters, too – just ask Leo Baekeland's mum, or Grace Hopper's dad. Here, again, we have reasons to be optimistic. There's probably a lot more we could do to improve schooling through technology: indeed, that's a plausible candidate for future economy-changing inventions. But already any child in an urban slum with an internet connection has greater potential access to knowledge than I had in a university library in the 1990s.

Other lessons seem easier to forget, like the value of allowing smart people to indulge their intellectual curiosity without a clear idea of where it might lead. In bygone days, this implied a wealthy man like Leo Baekeland tinkering in his lab; in the more recent past, it's meant government funding for basic research – producing the kind of technologies that enabled Steve Jobs and his team to invent the iPhone. Yet basic research is inherently unpredictable. It can take decades before anybody makes money putting into action what's been learned. That makes it a tough sell to private investors, and an easy target for government cutbacks in times of austerity.

Sometimes inventions just bubble up without any particular use in mind – the laser is a famous example, and paper was originally used for wrapping, not writing. But many of the inventions we've encountered have resulted from efforts to solve a specific problem, from Willis Carrier's air conditioning to Frederick McKinley Jones's refrigerated truck. That suggests that if we want to encourage more good ideas, we can concentrate minds by offering prizes for problem solving. Remember how the Longitude prize inspired Harrison to create his remarkable clocks?

There's recently been fresh interest in this idea: for example, the DARPA Grand Challenge, which began in 2004, helped kick-start progress in self-driving cars; on the three-hundredth anniversary of the original Longitude prize, the UK's innovation agency Nesta launched a new 'Longitude prize' for progress in testing microbial resistance to antibiotics; perhaps the biggest prize of all is the 'pneumococcal advanced market commitment' which has rewarded the development of vaccines with a $1.5 billion pot, supplied by five donor governments and the Gates Foundation.

The promise of profit is a constant motivator, of course. And we've seen how intellectual property rights can add credibility to that promise, by rewarding the successful inventor with a time-limited monopoly. But we also saw that this is a double-edged sword, and there's an apparently inexorable trend towards making intellectual property rights even longer and broader despite a widespread view among economists that they're already so overreaching that they're strangling innovation.

More broadly, what kind of laws and regulations encourage innovation is a question with no easy answers. The natural assumption is that bureaucrats should err on the side of getting out of the way of inventors, and we've seen this pay dividends. A laissez-faire approach gave us M-Pesa. But it also gave us the slow-motion disaster of leaded petrol; there are some inventions that governments really should be stepping in to prevent. And the process that produced the technology inside the iPhone was anything but laissez-faire.

Some hotbeds of research and development, like medicine, have well-established governance structures that are arguably sometimes too cautious. In other areas, from space to cyberspace, regulators are scrambling to catch up. And it's not only premature or heavy-handed regulation that can undermine the development of an emerging technology – so, paradoxically, can a total lack of regulation. If you're investing in drones, say, you want reassurance that irresponsible competitors won't be able to rush their half-ready products to market, creating a spate of accidents and a public backlash that causes governments to ban the technology altogether.

Regulators' task is complicated because, as we saw with

public key cryptography, most inventions can be used for either good or ill. How to manage the risks of 'dual use' technologies could become an increasingly vexed dilemma: only big states can afford nuclear missile programmes, but soon almost anyone might be able to afford a home laboratory that could genetically engineer bacteriological weapons – or innovative new medicines.

Adding to the challenge, the potential of inventions often becomes clear only when they combine with other inventions: think of the elevator, air conditioning and reinforced concrete, which together gave us the skyscraper. Now imagine combining a hobbyist's quadrocopter drone, facial recognition and geolocation software, and a 3D printer with a digital template for a gun: you have, hey presto, a home-made autonomous assassination drone. How are we supposed to anticipate the countless possible ways future inventions will interact? It's easy to demand that our politicians just get it right – but starry-eyed to expect that they will.

However, perhaps the biggest challenge that future inventions will create for governments is that new ideas tend to create losers as well as winners. Often, we regard that as just tough luck: nobody clamoured for compensation for second-tier professional musicians whose work dried up because of the gramophone; nor did the barcode and shipping container lead to subsidies for Mom-and-Pop shops to keep their prices competitive with Wal-Mart.

But when the losers are a wide enough swathe of the population, the impact can be socially and politically tumultuous. The industrial revolution ultimately raised living standards beyond what anyone might have dreamed in the eighteenth century, but it took the military to subdue the Luddites, who correctly perceived that it was disastrous for them. It's hardly fanciful to see echoes of Ned Ludd in the electoral surprises of 2016, from Brexit to President Trump. The technologies that enabled globalisation have helped lift millions out of poverty in countries like China – one of the poorest places on earth just fifty years ago, and now a solidly middle-income economy – but left whole communities in post-industrial regions of Western countries struggling to find new sources of stable, well-paid employment.

While populists surf the wave of anger by blaming immigrants and free trade, bigger long-term pressures always come from technological change. What will President Trump do if – when – self-driving vehicles replace 1.9 million American truck drivers? He doesn't have an answer; few politicians do.

We've already discussed one possible approach: a universal basic income, payable to all citizens. That's the sort of radical thinking we might need if artificial intelligence and robots really do live up to the hype and start outperforming humans at any job you care to name. But, like any new idea, it would cause new problems: notably, who's in and who's out? The welfare state and the passport prop each other up – and while universal basic income is an attractive idea in some ways, it looks less utopian when combined with impenetrable border walls.

In any case, my guess is that worries about the robot job-apocalypse are premature. It's very much in our minds now, but a final lesson from our fifty inventions is not to get too dazzled by the hottest new thing: in 2006, for example, MySpace surpassed Google as the most visited website in the United States; today, it doesn't make the top thousand. Writing in 1967, Kahn and Wiener made grand claims for the future of the fax machine. They weren't entirely wrong – but the fax machine is now close to being a museum piece.

Many of the inventions we've considered in these pages are neither new nor especially sophisticated, starting with the plough: it's no longer the technological centre of our civilisation, but it's still important and its design has changed less than we might think. This old technology still works and it still matters.

This isn't just a call for us to appreciate the value of old ideas, although it is partly that: an alien engineer visiting from Alpha Centauri might suggest it would be good if the enthusiasm we had for flashy new things was equally expressed for fitting more S-bends and pouring more concrete floors.

It's also a reminder that systems have their own inertia, an idea we encountered with Rudolf Diesel's engine: once fossil-fuelled internal combustion engines reached a critical mass, good luck with popularising peanut oil or getting investors to fund research into improving the steam engine. Some systems work so well, it's hard to think why anyone would want to rethink them – like the shipping container. But even systems that most people agree could have been done better, like the QWERTY keyboard layout, are remarkably resistant to change.

So bad decisions cast a long shadow. But the benefits of good decisions can last a surprisingly long time. And, for all the unintended consequences and unwelcome side effects of the inventions we've considered in these pages, overall they've had vastly more good effects than bad.

Sometimes, as our last invention will show, they've improved our lives almost beyond our ability to measure.

EPILOGUE: 50

The Light Bulb

Back in the mid-1990s the economist William Nordhaus conducted a series of simple experiments. One day, for example, he used a prehistoric technology: he lit a wood fire. Humans have been gathering and chopping and burning wood for tens of thousands of years. But Nordhaus also had a piece of high-tech equipment with him: a Minolta light meter. He burned 20 pounds of wood, kept track of how long it burned for and carefully recorded the dim, flickering firelight with his meter.

Another day, Nordhaus bought a Roman oil lamp – a genuine antique, he was assured – fitted it with a wick, and filled it with cold-pressed sesame oil. He lit the lamp and watched the oil burn down, again using the light meter to measure its soft, even glow. Nordhaus's open wood fire had burned for just three hours when fuelled with 20 pounds of wood. But a mere eggcup of oil burned all day, and more brightly and controllably.

Why was Nordhaus doing this? He wanted to understand the economic significance of the light bulb. But that was just part of a larger project. Nordhaus wanted, if you'll forgive the play on words, to shed light on a difficult issue for economists: how to keep track of inflation, of the changing cost of goods and services.

To see why this is difficult, consider the price of travelling from – say – Lisbon in Portugal to Luanda in Angola. When the journey was first made by Portuguese explorers, it would have been an epic expedition, taking months. Later, by steam ship, it would have taken a few days. Then by plane, a few hours. An economic historian who wanted to measure inflation could start by tracking the price of passage on the steamer. But then, once an air route opens up, which price do you look at? Perhaps you simply switch to the airline ticket price once more people start flying than sailing. But flying is a different service – faster, more convenient. If more travellers are willing to pay twice as much to fly, it hardly makes sense for inflation statistics to record that the cost of the journey has suddenly doubled. How, then, do we measure inflation when what we're able to buy changes so radically over time?

This question isn't merely a technical curiosity. How we answer it underpins our view of human progress over the centuries. The economist Timothy Taylor begins his introductory economics lectures by asking his students: would they rather be making 70,000 dollars a year now, or 70,000 dollars in 1900?

At first glance that's a no-brainer. Seventy thousand dollars in the year 1900 is a much better deal. It's worth about two million dollars in today's terms, after adjusting for inflation. A dollar would buy much more in 1900 – enough steak to feed a family; enough bread for a fortnight. For a dollar you could hire a man to work for you all day. Your 70,000-dollar salary would easily pay for a mansion, housemaids and a butler.

Yet of course, in another way a dollar in 1900 buys much less than today. A dollar today will buy you an international phone call on a portable phone, or a day's worth of broadband internet access – or a course of antibiotics. In 1900, none of that was available, not to the richest men in the world.

And all this may explain why the majority of Timothy Taylor's students said they'd rather have a decent income now than a fortune a century ago. And it's not just the high-tech stuff; they also knew their money would buy better central heating, better air conditioning and a much better car – even if they had no butler and fewer steak dinners. Inflation statistics tell us that 70,000 dollars today is worth much less than the same amount in 1900. But people who've experienced modern technology don't see things that way.

Because we don't have a good way to compare an iPod today to a gramophone a century ago, we don't really have a good way to quantify how much all the inventions described in this book have really expanded the choices available to us. We probably never will.

But we can try – and Bill Nordhaus was trying as he fooled around with wood fires, antique oil lamps and Minolta light meters. He wanted to unbundle the cost of a single quality that humans have cared deeply about since time immemorial, using the state-of-the-art technology of different ages: illumination. That's measured in lumens, or lumen-hours. A candle, for example, gives off 13 lumens while it burns; a typical modern light bulb is almost a hundred times as bright as that.

Imagine a hard week's work gathering and chopping wood, ten hours a day for six days. Those sixty hours of work would produce 1000 lumen hours of light. That's the equivalent of one modern light bulb shining for just fifty-four minutes, although what you'd actually get is many more hours of dim, flickering light instead. Of course, light isn't the only reason to burn fires: keeping warm, cooking food and frightening off wild animals are also benefits. Still, if you wanted light and a wood fire was your only option, you might instead decide to wait until the sun came up before doing what you wanted.

Thousands of years ago, better options came along – candles from Egypt and Crete, and oil lamps from Babylon. The light they provided was steadier and more controllable, but still prohibitively expensive. In a diary entry of May 1743, the president of Harvard University, Reverend Holyoake, noted that his household had spent two days making 78 pounds of tallow candles. Six months later he noted, stenographically, 'Candles all gone.' And those were the summer months.

Nor were these the romantic, clean-burning paraffin wax candles we use today. The wealthiest people could afford beeswax, but most people – even the Harvard president – used tallow candles: stinking, smoking sticks of animal fat. Making them involved heating up the animal fat and patiently dipping and re-dipping wicks into the molten lard. It was pungent and time-consuming work. According to Nordhaus's research, if you set aside one whole week a year to spend sixty hours devoted exclusively to making candles – or earning the money to buy them – that would enable you to burn a single candle for just two hours and twenty minutes every evening.

Things improved a little as the eighteenth and nineteenth centuries unfolded. Candles were made of spermaceti – the milk-hued oily gloop harvested from dead sperm whales. Ben Franklin loved the strong, white light they gave off, and the way they 'may be held in the Hand, even in hot Weather, without softening; that their Drops do not make Grease Spots like those from common Candles; that they last much longer'. While the new candles were pleasing, they were pricey. George Washington calculated that burning a single spermaceti candle for five hours a night all year would cost him £8 – that is well over a thousand dollars in today's money. A few decades later, gas lamps and kerosene lamps helped to lower lighting costs; they also saved the sperm whale from extinction. But they, too, were basically an expensive hassle. They tipped, dripped, smelt and set fire to things.

Then something changed. That something was the light bulb.

By 1900, one of Thomas Edison's carbon filament bulbs would provide you with ten days of continuous illumination, a hundred times as bright as a candle, for the money you'd earn with our sixty-hour week of hard labour. By 1920, that same week of labour would pay for more than five months' continuous light from tungsten filament bulbs; by 1990, it was ten years. A couple of years after that, thanks to compact fluorescent bulbs, it was more than five times longer. The labour that had once produced the equivalent of fifty-four minutes of quality light now produced fifty-two years. And modern LED lights continue to get cheaper and cheaper.

Switch off a light bulb for an hour and you're saving illumination that would have cost our ancestors all week to create. It would have taken Benjamin Franklin's contemporaries all afternoon. But someone in a rich industrial economy today could earn the money to buy that illumination in a fraction of a second. And of course light bulbs are clean, fire-safe and controllable – no flicker, or stink of pig fat, or risk of fire. You could leave a child alone with one.

None of this has been reflected in traditional measures of inflation, which Nordhaus reckons has overstated the price of light by a factor of about a thousand since 1800. Light seems to have become more expensive over time, but in fact it's vastly cheaper. Timothy Taylor's students instinctively feel that they could buy more of what they really want with 70,000 dollars today than with that amount in 1900. Nordhaus's investigations suggest that – when it comes to light, at least – they're quite right.

That's why I wanted to finish this book with the story of light. Not the now familiar development of the incandescent bulb by Thomas Edison and Joseph Swan – but the story of how, over the centuries, humanity has developed innovation after innovation to utterly revolutionise our access to light.

Those innovations have transformed our society into one where we can work whenever we want to work, to read or to sew or to play whenever we want to play, regardless of how dark the night has become.

No wonder the light bulb is still the visual cliché for 'new idea' – literally, the icon for invention. Yet even iconic status underrates it. Nordhaus's research suggests that however much we venerate it, perhaps we do not venerate it enough. The price of light alone tells that story: it has fallen by a factor of five hundred thousand, far faster than official statistics suggest, and so fast that our intuition cannot really grasp the miracle of it all.

Man-made light was once a thing that was too precious to use. Now it is too cheap to notice. If ever there was a reminder that progress is possible – that amid all the troubles and challenges of modern life, we have much to be grateful for – then this is it.

Notes

1 The Plough

1 *No internet. No electricity. No fuel.* For a deep exploration of this scenario, see Lewis Dartnell, *The Knowledge: How to rebuild our world after an apocalypse* (London: Vintage, 2015).

1 *a simple yet transformative technology* James Burke, BBC TV documentary *Connections* (1978a).

2 *and people followed* James Burke, *Connections* (London: Macmillan, 1978b), p. 7; Ian Morris, *Foragers, Farmers and Fossil Fuels* (Oxford: Princeton University Press, 2015).

2 *it happened almost everywhere* Morris, p. 153.

2 *the foragers they had replaced* Morris, p. 52. Morris uses consumption of energy (in food and other forms) as his measure of income – reductive, but given that we're talking about prehistory here, not unreasonable.

3 *making cities, building civilisation* Burke 1978a. In *The Economy of Cities* (New York: Vintage, 1970) Jane Jacobs sets out an alternative view: the city came first, in the form of a trading settlement that gradually became something more complex and permanent. Only then came agricultural technologies such as the plough and domesticated animals and crops. Either way, the plough came early in civilisation, and has been essential ever since.

3 *would simply have starved* Branko Milanovic, Peter H. Lindert, Jeffrey G. Williamson, 'Measuring Ancient Inequality', NBER Working Paper no. 13550, October 2007.

3 *turns it upside down* Dartnell, pp. 60–2.

4 *the manorial system in northern Europe* Lynn White, *Medieval Technology and Social Change* (Oxford: Oxford University Press, 1962), pp. 39–57.

4 *kneeling, twisting and grinding grain* Morris, p. 59.

4 *more frequent pregnancies* Jared Diamond, 'The Worst Mistake in the History of the Human Race', *Discover*, May 1987, http://discovermagazine.com/1987/may/02-the-worst-mistake-in-the-history-of-the-human-race.

5 *it's been slow to fade* Morris, p. 60.

6 *so many mongongo nuts in the world* Diamond, ibid.

INTRODUCTION

7   *Shea Terra Organics Company* https://www.evitamins.com/uk/mongongo-hair-oil-shea-terra-organics-108013, accessed 17 January 2017.

7   *major economic centres* This is an educated guess courtesy of Eric Beinhocker, director of the Institute for New Economic Thinking at Oxford University.

I. WINNERS AND LOSERS

13 *the fabric and clothing industries* Walter Isaacson, 'Luddites fear humanity will make short work of finite wants', *Financial Times*, 3 March 2015, https://www.ft.com/content/9e9b7134-c1a0-11e4-bd24-00144feab7de.

13 *right to dread it* Tim Harford, 'Man vs Machine (Again)', *Financial Times*, 13 March 2015, https://www.ft.com/content/f1b39a64-c762-11e4-8e1f-00144feab7de; Clive Thompson, 'When Robots Take All of Our Jobs, Remember the Luddites', *Smithsonian Magazine*, January 2017, http://www.smithsonianmag.com/innovation/when-robots-take-jobs-remember-luddites-180961423/.

14 *the machines would empower* Evan Andrews, 'Who Were the Luddites?' *History*, 7 August 2015, http://www.history.com/news/ask-history/who-were-the-luddites.

2 The Gramophone

15 *only one Elton John* 'The World's 25 Highest-Paid Musicians', *Forbes* http://www.forbes.com/pictures/eegi45lfkk/the-worlds-25-highest-paid-musicians/.

15 *Mrs Billington singing* 'Mrs Billington, as St Cecilia', *British Museum Collection*, http://www.britishmuseum.org/research/collection_online/collection_object_details.aspx?objectId=1597608&partId=1; Chrystia Freeland, 'What a Nineteenth-Century English Soprano Can Teach Us About the Income Gap', *Penguin Press Blog*, 1 April 2013, http://thepenguinpress.com/2013/04/elizabeth-billington/.

16 *illuminated for three days* W.B. Squire, 'Elizabeth Billington', *The Dictionary of National Biography 1895–1900*, https://en.wikisource.org/wiki/Billington,_Elizabeth_(DNB00).

16 *'than ever before'* Alfred Marshall, *Principles of Economics*, 1890, cited in Sherwin Rosen, 'The Economics of Superstars', *American Economic Review* Vol. 71.5, December 1981.

17 *back into sound again* 'Oldest Recorded Voices Sing Again', *BBC News*, 28 March 2008, http://news.bbc.co.uk/1/hi/technology/7318180.stm.

17 *a mere 200 records* Tim Brooks, *Lost Sounds: Blacks and the Birth of the Recording Industry, 1890–1919* (Chicago: University of Illinois Press, 2004), p. 35.

17 *Alfred Marshall had described* Richard Osborne, *Vinyl: A History of the Analogue Record* (Farnham: Ashgate, 2012).

18 *phonographs in 1801* Sherwin Rosen, 'The Economics of Superstars', *American Economic Review* Vol. 71.5, December 1981.

18 *two divisions below* 'Mind the Gap', *Daily Mail*, 20 February 2016, http://www.dailymail.co.uk/sport/football/article-3456453/Mind-gap-Premier-League-wages-soar-average-salaries-2014-15-season-1-7million-rest-creep-along.html.

19 *'that's going to be left'* Cited in Alan Krueger, 'The Economics of Real Superstars: The Market for Rock Concerts in the Material World', working paper, April 2004.

19 *95 per cent put together* Alan B. Krueger, 'Land of Hope and Dreams: Rock and Roll, Economics and Rebuilding the Middle Class', speech on 12 June 2013 in Cleveland, OH, https://obamawhitehouse.archives.gov/blog/2013/06/12/rock-and-roll-economics-and-rebuilding-middle-class.

3 Barbed Wire

20 *But never mind* Alan Krell, *The Devil's Rope: A Cultural History of Barbed Wire* (London: Reaktion Books, 2002), p. 27.

21 *'cheaper than dust'* Ian Marchant, *The Devil's Rope*, BBC Radio 4 documentary, http://www.bbc.co.uk/programmes/b048l0s1, Monday 19 January 2015.

21 *stop the barbs from sliding around* Olivier Razac, *Barbed Wire: A political history*, English translation by Jonathan Kneight (London: Profile Books, 2002).

21 *the future of the nation* http://www.historynet.com/homestead-act and http://plainshumanities.unl.edu/encyclopedia/doc/egp.ag.011.

22 *'The American Desert'* See Joanne Liu's map at the 99% Invisible website: http://99percentinvisible.org/episode/devils-rope/.

22 *settle the American West* 'The Devil's Rope', 99% Invisible Episode 157, 17 March 2015, http://99percentinvisible.org/episode/devils-rope/.

22 *rest of the world put together* Razac, pp. 5–6.

23 *fights started to break out* Texas State Historical Association, 'Fence Cutting', https://www.tshaonline.org/handbook/online/articles/auf01.

23 *'getting ripe for market'* Alex E. Sweet and J. Armoy Knox, *On an American Mustang, Through Texas, From the Gulf to the Rio Grande*, 1883, https://archive.org/stream/onmexicanmustang00swee/onmexicanmustang00swee_djvu.txt.

24 *'the lands of America now'* Barbara Arneil, 'All the World Was America', doctoral thesis, University College London, 1992, http://discovery.ucl.ac.uk/1317765/1/283910.pdf.

25 *digital barbed wire* Cory Doctorow, 'Lockdown: The Coming War on General-purpose Computing', http://boingboing.net/2012/01/10/lockdown.html; 'Reply All #90: Matt Lieber Goes To Dinner', https://gimletmedia.com/episode/90-matt-lieber-goes-to-dinner/.

25 *ten times over* Marchant, ibid.

4 Seller Feedback

26 *a ride with himself* http://www.bloomberg.com/news/articles/2015-06-28/one-driver-explains-how-he-is-helping-to-rip-off-uber-in-china.

28 *he could fix it up* https://www.ebayinc.com/stories/news/meet-the-buyer-of-the-broken-laser-pointer/.

28 *it fell out of fashion* http://www.socresonline.org.uk/6/3/chesters.html.

29 *'would have grown without it'* https://player.vimeo.com/video/130787986.

30 *to suffer as a result* Tim Harford, 'From Airbnb to eBay, the best ways to combat bias', *Financial Times*, 16 November 2016, https://www.ft.com/content/7a170330-ab84-11e6-9cb3-bb8207902122 and Benjamin G. Edelman, Michael Luca, and Daniel Svirsky, 'Racial Discrimination in the Sharing Economy: Evidence from a Field Experiment', *American Economic Journal: Applied Economics* (forthcoming).

5 Google Search

31 *Lancaster University* http://www.bbc.co.uk/news/magazine-36131495.

32 *filled with porn websites* John Battelle, *The Search: How Google and Its Rivals Rewrote the Rules of Business and Transformed Our Culture* (London: Nicholas Brealey Publishing, 2006).

32 *better way to search the web* Battelle, p. 78.

33 *tens of billions of dollars* http://www.statista.com/statistics/266472/googles-net-income/.

33 *viable business models* Battelle, Chapter 5.

33 *fallen off a cliff* https://www.techdirt.com/articles/20120916/14454920395/newspaper-ad-revenue-fell-off-quite-cliff-now-par-with-1950-revenue.shtml.

33 *to list the most important* 'The impact of Internet technologies: Search', July 2011, McKinsey https://www.mckinsey.com/~/media/McKinsey/dotcom/client_service/High%20Tech/PDFs/Impact_of_Internet_technologies_search_final2.ashx.

34 *an unanticipated complication* https://www.nytimes.com/2016/01/31/business/fake-online-locksmiths-may-be-out-to-pick-your-pocket-too.html.

35 *stamp out this sort of thing* See 'Reply All #76: Lost In A Cab', https://gimletmedia.com/episode/76-lost-in-a-cab/.

35 *organic search results* http://www.statista.com/statistics/216573/worldwide-market-share-of-search-engines/.

35 *'you can only guess'* http://seo2.0.onreact.com/10-things-the-unnatural-links-penalty-taught-me-about-google-and-seo.

36 *searched for before* https://hbr.org/2015/03/data-monopolists-like-google-are-threatening-the-economy.

6 Passports

37 *'our national privileges'* Martin Lloyd, *The Passport: The history of man's most travelled document* (Canterbury: Queen Anne's Fan, 2008), p. 63.

37 *back to biblical times* Craig Robertson, *The Passport in America: The History of a Document* (Oxford: Oxford University Press, 2010), p. 3.

38 *the relevant government minister* Lloyd, p. 200.

38 *skilled workers from leaving* Ibid, p. 3.

38 *abolished them in 1860* Ibid, pp. 18 and 95.

38 *at least in peacetime* Ibid, pp. 18, 95–6.

38 *if you were white* Jane Doulman, David Lee, *Every Assistance and Protection: A History of the Australian Passport* (Sydney: Federation Press, 2008), p. 34.

38 *in the constitution* Lloyd, p. 95.

38 *to venture inland* Ibid, pp. 70–1.

38 *might soon disappear altogether* Ibid, pp. 96–7.

38 *Bodrum, Turkey* http://time.com/4162306/ alan-kurdi-syria-drowned-boy-refugee-crisis/.

39 *plane tickets for them all* http://www.independent.co.uk/news/world/europe/ aylan-kurdi-s-story-how-a-small-syrian-child-came-to-be-washed-up-on-a-beach-in-turkey-10484588.html.

39 *the Kurdis had no passports* http://www.bbc.co.uk/news/world-europe-34141716.

39 *they'd have had no problems* http://www.cic.gc.ca/english/visit/visas-all.asp.

39 *St Kitts and Nevis* http://www.bbc.co.uk/news/business-27674135.

40 *should not be let in* http://www.independent.co.uk/news/world/europe/ six-out-of-10-migrants-to-europe-come-for-economic-reasons-and-are-not-refugees-eu-vice-president-a6836306.html.

40 *the arrival of immigrants* Amandine Aubrya, Michał Burzyńskia, Frédéric Docquiera, 'The Welfare Impact of Global Migration in OECD Countries', *Journal of International Economics*, 101 (2016), http://www.sciencedirect.com/ science/article/pii/S002219961630040X.

41 *to look for work* Mr Tebbit was actually relating a story about his own father's search for work. But most people inferred that he was telling jobless people in general to get on their bikes. http://news.bbc.co.uk/1/hi/programmes/ politics_show/6660723.stm.

41 *it would double* http://openborders.info/double-world-gdp/.

41 *with a photo* Lloyd, pp. 97–101.

7 Robots

42 *extracts a bottle* https://www.youtube.com/watch?v=aA12i3ODFyM.

42 *selling it by 2020* http://spectrum.ieee.org/automaton/robotics/industrial-robots/ hitachi-developing-dual-armed-robot-for-warehouse-picking.

42 *to pick things off* https://www.technologyreview.com/s/538601/inside-amazons-warehouse-human-robot-symbiosis/.

42 *up to fourfold* http://news.nationalgeographic.com/2015/06/150603-science-technology-robots-economics-unemployment-automation-ngbooktalk/.

43 *tasks like welding* http://www.robotics.org/joseph-engelberger/unimate.cfm.

43 *showing it what to do* http://newatlas.com/baxter-industrial-robot-positioning-system/34561/.

43 *doubling every five years* http://www.ifr.org/news/ifr-press-release/world-robotics-report-2016-832/.

43 *robots are part of that* http://foreignpolicy.com/2014/03/28/made-in-the-u-s-a-again/.

43 *lettuce-pickers* http://www.techinsider.io/companies-that-use-robots-instead-of-humans-2016-2/#quiet-logistics-robots-quickly-find-package-and-ship-online-orders-in-warehouses-2.

43 *bartenders* http://www.marketwatch.com/story/9-jobs-robots-already-do-better-than-you-2014-01-27.

43 *hospital porters* https://www.wired.com/2015/02/incredible-hospital-robot-saving-lives-also-hate/.

43 *despite recent progress* http://fortune.com/2016/06/24/rosie-the-robot-data-sheet/.

44 *its sense of balance* https://www.weforum.org/agenda/2015/04/qa-the-future-of-sense-and-avoid-drones.

44 *a bold prediction* Nick Bostrom, *Superintelligence: Paths, Dangers, Strategies* (Oxford: Oxford University Press, 2014).

45 *books on economics* http://fortune.com/2015/02/25/5-jobs-that-robots-already-are-taking/.

45 *wages are stagnating* https://www.technologyreview.com/s/515926/how-technology-is-destroying-jobs/.

45 *zero or below* Klaus Schwab, *The Fourth Industrial Revolution* (World Economic Forum, 2016).

46 *they still can't clean toilets* http://news.nationalgeographic.com/2015/06/150603-science-technology-robots-economics-unemployment-automation-ngbooktalk/.

46 *'pick 19'* https://www.ft.com/content/da557b66-b09c-11e5-993b-c425a3d2b65a.

8 The Welfare State

47 *persuading men to accept her ideas* http://www.nytimes.com/2006/02/12/books/review/women-warriors.html.

47 *pensions for the elderly* Kirstin Downey, *The Woman Behind the New Deal: The Life and Legacy of Frances Perkins—Social Security, Unemployment Insurance, and the Minimum Wage* (New York: Anchor Books, 2010).

48 *discouraging them from working* http://www.cato.org/publications/policy-analysis/work-versus-welfare-trade-europe.

49 *grandmothers started getting pensions* http://economics.mit.edu/files/732.

49 *positive and negative effects balance out* https://inclusivegrowth.be/visiting-grants/outputvisitis/c01-06-paper.pdf.

49 *each individual's slice* Lane Kenworthy, *Do social welfare policies reduce poverty? A cross-national assessment*, East Carolina University https://lanekenworthy.files.wordpress.com/2014/07/1999sf-poverty.pdf.

49 *haven't been doing that so well* Koen Caminada, Kees Goudswaard and Chen Wang, *Disentangling Income Inequality and the Redistributive Effect of Taxes*

and Transfers in 20 LIS Countries Over Time, September 2012, http://www.
lisdatacenter.org/wps/liswps/581.pdf.

49   *may widen further* In the UK, for example, see the Institute for Fiscal Studies
     presentation on *Living Standards, Poverty and Inequality 2016*, https://www.
     ifs.org.uk/uploads/publications/conferences/hbai2016/ahood_income%20
     inequality2016.pdf. The World Wealth and Incomes Database http://www.
     wid.world/ collects together data on the income share of the top 10 per cent
     and top 1 per cent in various countries.

49   *full-time and long-lasting* https://www.chathamhouse.org/sites/files/chathamhouse/
     field/field_document/20150917WelfareStateEuropeNiblettBeggMushovel.pdf.

50   *a self-employed builder will not* Benedict Dellot and Howard Reed,
     Boosting the living standards of the self-employed, RSA, March 2015,
     https://www.thersa.org/discover/publications-and-articles/reports/
     boosting-the-living-standards-of-the-self-employed.

50   *on the path to Brexit* http://www.telegraph.co.uk/news/2016/05/19/
     eu-deal-what-david-cameron-asked-for-and-what-he-actually-got/.

50   *Karl Marx and Friedrich Engels* https://www.dissentmagazine.org/
     online_articles/bruce-bartlett-conservative-case-for-welfare-state.

50   *his socialist opponents* M. Clark, *Mussolini* (London: Routledge 2014).

51   *a universal basic income* http://www.ft.com/cms/s/0/7c7ba87e-229f-11e6-
     9d4d-c11776a5124d.html.

52   *hardly anyone gave up work* Evelyn L. Forget, *The Town With No Poverty: Using
     Health Administration Data to Revisit Outcomes of a Canadian Guaranteed Annual
     Income Field Experiment*, University of Manitoba, February 2011, https://
     public.econ.duke.edu/~erw/197/forget-cea%20(2).pdf.

52   *the same thing happens elsewhere* http://www.ft.com/cms/s/0/7c7ba87e-229f-
     11e6-9d4d-c11776a5124d.html.

52   *the entire federal budget* http://www.bloomberg.com/view/articles/2016-06-06/
     universal-basic-income-is-ahead-of-its-time-to-say-the-least.

52   *across the nation* http://www.newyorker.com/magazine/2016/06/20/
     why-dont-we-have-universal-basic-income.

II REINVENTING HOW WE LIVE

53   *a concealed zip-fastener* Luke Lewis, '17 Majestically Useless Items from the
     Innovations Catalogue' *Buzzfeed* https://www.buzzfeed.com/lukelewis/
     majestically-useless-items-from-the-innovations-catalogue?utm_term=.
     rjJpZjxz4Y#.fjJXKxp6y7

9 Infant Formula

55   *four thousand feet shorter* http://www.scientificamerican.com/
     article/1816-the-year-without-summer-excerpt/.

55   *rats, cats and grass* http://jn.nutrition.org/content/132/7/2092S.full.

56   *polishes for sale* William H. Brock, *Justus von Liebig: The Chemical Gatekeeper,*

Cambridge Science Biographies (Cambridge: Cambridge University Press, 2002).

56 *fats, proteins and carbohydrates* Harvey A. Levenstein, *Revolution at the Table: The Transformation of the American Diet* (Berkeley: University of California Press, 2003).

56 *beef extract* http://www.ft.com/cms/s/2/6a6660e6-e88a-11e1-8ffc-00144feab49a.html.

56 *rigorous scientific study* http://www.ncbi.nlm.nih.gov/pmc/articles/PMC2684040/.

56 *killed the mother* http://ajcn.nutrition.org/content/72/1/241s.full.pdf.

56 *in the poorest countries today* http://data.worldbank.org/indicator/SH.STA.MMRT.

56 *one in twenty* Marianne R. Neifert, *Dr. Mom's Guide to Breastfeeding* (New York: Plume, 1998).

56 *Liebig's invention* http://www.ncbi.nlm.nih.gov/pmc/articles/PMC2684040/.

56 *teemed with bacteria* http://www.ncbi.nlm.nih.gov/pmc/articles/PMC2379896/pdf/canfamphys00115-0164.pdf.

56 *their first birthday* Geoff Talbot, *Specialty Oils and Fats in Food and Nutrition: Properties, Processing and Applications* (Woodhead Publishing, 2015), p. 287.

56 *their working patterns* Marianne Bertrand, Claudia Goldin and Lawrence F. Katz, 'Dynamics of the Gender Gap for Young Professionals in the Financial and Corporate Sectors', *American Economic Journal: Applied Economics* 2(3), 2010, 228–55.

57 *legal right to take time off* https://www.theguardian.com/money/shortcuts/2013/nov/29/parental-leave-rights-around-world.

57 *encouraging them to take it* https://www.washingtonpost.com/news/on-leadership/wp/2015/11/23/why-mark-zuckerberg-taking-paternity-leave-really-matters/?utm_term=.c36a3cbfe8c0.

57 *persevere with breastfeeding* http://www.ncbi.nlm.nih.gov/pmc/articles/PMC3387873/.

58 *800,000 children's lives each year* http://www.who.int/pmnch/media/news/2016/lancet_breastfeeding_partner_release.pdf?ua=1.

58 *300 billion dollars a year* http://www.who.int/pmnch/media/news/2016/lancet_breastfeeding_partner_release.pdf?ua=1.

58 *the global formula market* http://www.slideshare.net/Euromonitor/market-oveview-identifying-new-trends-and-opportunities-in-the-global-infant-formula-market.

59 *the advertising swayed* Levenstein, ibid.

59 *widely flouted* http://www.businessinsider.com/nestles-infant-formula-scandal-2012-6?IR=T#the-baby-killer-blew-the-lid-off-the-formula-industry-in-1974-1.

59 *some died* BBC News 'Timeline: China Milk Scandal' 25 January 2010, http://news.bbc.co.uk/1/hi/7720404.stm.

59 *a hundred dollars a litre* http://www.sltrib.com/news/3340606-155/got-breast-milk-if-not-a.

10 TV Dinners

60 *many hours each week* Alison Wolf, *The XX Factor* (London: Profile Books, 2013), pp. 80–5.

61 *'That was enough'* Jim Gladstone, 'Celebrating (?) 35 years of TV dinners', *Philly.com*, 2 November 1989, http://articles.philly.com/1989-11-02/entertainment/26137683_1_tv-dinner-frozen-dinner-clarke-swanson.

62 *in the 1960s* USDA, http://www.ers.usda.gov/topics/food-choices-health/food-consumption-demand/food-away-from-home.aspx.

62 *at grocery stores* Matt Philips, 'No One Cooks Any More', *Quartz*, 14 June 2016, http://qz.com/706550/no-one-cooks-anymore/.

62 *over a decade before* Wolf, p. 83.

62 *macaroni in the 1880s* Ruth Schwartz Cowan, *More Work for Mother* (London: Free Association Books, 1989), pp. 48–9 and especially pp. 72-3. (Professor Cowan also offers a touching postscript about her own experience with laundry.)

63 *housework that women did* Wolf, p. 84 and Valerie Ramey, 'Time spent in home production in the 20th century', NBER Paper 13985 (2008); Valerie Ramey, 'A century of work and leisure', NBER Paper 12264 (2006).

63 *willing to starve* Wolf, p. 85.

64 *in terms of time* David Cutler, Edward Glaeser and Jesse Shapiro, 'Why have Americans become more obese?' *Journal of Economic Perspectives* (2003) 17, no. 3: 93–118; doi:10.1257/089533003769204371.

11 The Pill

66 *a cervical cap* Jonathan Eig, *The Birth of the Pill* (London: Macmillan, 2014), p. 7.

67 *the diaphragm isn't much better* James Trussell, 'Contraceptive Failure in the United States', *Contraception*, 83(5), May 2011, pp. 397–404.

67 *women in the US* Unless otherwise stated, the argument and statistics here are from Claudia Goldin and Lawrence Katz, 'The Power of the Pill: Oral Contraceptives and Women's Career and Marriage Decisions', *Journal of Political Economy*, 110(4), 2002.

69 *so did women's wages* A later study by economist Martha Bailey used a similar state-by-state analysis to examine the impact of oral contraceptives on women's wages. She, too, found a major effect. Women who had had access to them between the ages of eighteen and twenty-one were making 8 per cent more than women who had not.

69 *before having children* Steven E. Landsburg, 'How much does motherhood cost?' *Slate*, 9 Dec 2005, http://www.slate.com/articles/arts/everyday_economics/2005/12/the_price_of_motherhood.html, and Amalia R. Miller, 'The effects of motherhood timing on career path', *Journal of Population Economics*, 24(3), July 2011, pp. 1071–1100, http://www.jstor.org/stable/41488341.

70 *a few months behind* Carl Djerassi, one of the fathers of the pill, has a chapter on the Japanese case in *This Man's Pill* (Oxford: Oxford University Press, 2001).

70 *recognition in the workplace* See for example the World Economic Forum
    Gender Gap report and ranking: https://www.weforum.org/reports/
    global-gender-gap-report-2015/ and http://www.japantimes.co.jp/
    news/2014/10/29/national/japan-remains-near-bottom-of-gender-gap-
    ranking/#.V0cFlJErI2w.

12 Video Games

71 *playing Spacewar* Steven Levy, *Hackers: Heroes of the Computer Revolution*
    (Cambridge: O'Reilly, 2010), p. 55.
72 *all but the highly trained* J.M. Graetz, 'The Origin of Spacewar', *Creative
    Computing*, Vol. 7, No. 8, August 1981.
72 *computers worked for the suits* In 1972 Stewart Brand wrote a prescient piece
    for *Rolling Stone*, 'Fanatic Life and Symbolic Death Among the Computer
    Bums', about how Spacewar would transform our relationship to computers.
    His brilliant opening lines: 'Ready or not, computers are coming to the
    people. That's good news, maybe the best since psychedelics.' http://www.
    wheels.org/spacewar/stone/rolling_stone.html. More recently Steven
    Johnson has argued that Brand's article was almost as influential as Spacewar
    itself, helping people to understand how computers had been unlocked
    and turned into compelling sources of entertainment and enrichment for
    everyone – not just grey corporate calculators: *Wonderland: How Play Made
    the Modern World* (New York: Riverhead, 2016).
73 *the night sky above Lowell* Graetz, ibid.
73 *rival the film industry for revenue* It's often said that video games take in more
    revenue than films, but this claim stacks up only if we take a broad view
    of video games – including spending on games consoles – and a narrow
    view of movies, excluding rentals, streaming and DVD sales. Nevertheless
    the video games earn a large and growing revenue. See http://www.
    gamesoundcon.com/single-post/2015/06/14/Video-Games-Bigger-than-
    the-Movies-Dont-be-so-certain for a useful discussion.
73 *Edward Castronova* Edward Castronova, 'Virtual Worlds: A First-Hand
    Account of Market and Society on the Cyberian Frontier', CESifo Working
    Paper no. 618, December 2001.
74 *virtual game characters* Vili Lehdonvirta, 'Geographies of Gold Farming',
    Oxford Internet Institute blog post 29 October 2014, http://cii.oii.ox.ac.
    uk/2014/10/29/geographies-of-gold-farming-new-research-on-the-third-
    party-gaming-services-industry/ – and Vili Lehdonvirta interview with the
    author, 9 December 2016.
74 *a game called Lineage* 'Virtual Gaming Worlds Overtake Namibia', BBC
    News, 19 August 2004, http://news.bbc.co.uk/1/hi/technology/3570224.
    stm and 'Virtual Kingdom Richer than Bulgaria', BBC News, 29 March
    2002, http://news.bbc.co.uk/1/hi/sci/tech/1899420.stm.
74 *within easy reach* Jane McGonigal, *Reality is Broken* (London: Vintage, 2011),
    p. 3. McGonigal's estimates included 183 million in the US, 105 million in
    India, 200 million in China and 100 million in Europe.

75 *far more appealing* Ana Swanson, 'Why amazing video games could be causing a big problem for America', *The Washington Post*, 23 September 2016, https://www.washingtonpost.com/news/wonk/wp/2016/09/23/why-amazing-video-games-could-be-causing-a-big-problem-for-america/.

13 Market Research

76 *'what the consumer wanted'* 'The merchandising of automobiles; an address to retailers by Charles Coolidge Parlin, Manager, Division of Commercial Research', The Curtis Publishing Company, 1915, http://babel.hathitrust.org/cgi/pt?id=wu.89097464051;view=1up;seq=1.

76 *half a million people* http://www.bls.gov/ooh/business-and-financial/market-research-analysts.htm/.

77 *'as long as it's black'* http://www.bbc.com/news/magazine-23990211.

77 *a magazine publisher* Douglas Ward, *A New Brand of Business: Charles Coolidge Parlin, Curtis Publishing Company, and the Origins of Market Research* (Philadelphia: Temple University Press, 2010).

77 *'one which would continue'* 'The merchandising of automobiles; an address to retailers by Charles Coolidge Parlin, Manager, Division of Commercial Research', The Curtis Publishing Company, 1915. http://babel.hathitrust.org/cgi/pt?id=wu.89097464051;view=1up;seq=1.

78 *cheap and durable* Tom Collins, *The Legendary Model T Ford: The Ultimate History of America's First Great Automobile* (Iola, WI: Krause Publications, 2007), pp. 78, 155.

78 *advertising budgets almost doubled* Mansel G. Blackford and Austin K. Kerr, *Business Enterprise in American History* (Houghton Mifflin, 1993).

78 *collected royalties* http://www.economist.com/node/1632004.

78 *the 'role model'* ibid.

78 *to manufacture desire* Blackford and Kerr, ibid.

79 *different shades of blue* http://www.nytimes.com/2009/03/01/business/01marissa.html.

79 *'we never imagined'* Geoffrey Miller, *Must Have: The Hidden Instincts Behind Everything We Buy* (London: Vintage, 2010).

80 *'it's a better car'* http://dwight-historical-society.org/Star_and_Herald_Images/1914_Star_and_Herald_images/019_0001.pdf.

14 Air Conditioning

81 *slow down the greenhouse effect* http://www.economist.com/node/17414216.

81 *particles into the sky* https://www.scientificamerican.com/article/rain-how-to-try-to-make-it-rain/

82 *carry the cooler air inside* http://content.time.com/time/nation/article/0,8599,2003081,00.html.

82 *an 'ice famine'* Steven Johnson, *How We Got to Now* (London: Particular Books, 2014).

83 *such as theatres* http://www.williscarrier.com/1903-1914.php.

83 *unpleasant smells* Bernard Nagengast, 'The First Century of Air
    Conditioning', *ASHRAE Journal*, February 1999, https://www.ashrae.org/
    File%20Library/docLib/Public/200362710047_326.pdf.

84 *40 per cent* http://www.theatlantic.com/technology/archive/2011/07/
    keepin-it-cool-how-the-air-conditioner-made-modern-america/241892/.

84 *elected Ronald Reagan* Johnson, ibid.

84 *half the world's air conditioning* http://content.time.com/time/nation/
    article/0,8599,2003081,00.html.

84 *the global leader* https://www.theguardian.com/environment/2012/jul/10/
    climate-heat-world-air-conditioning.

84 *in just ten years* http://www.economist.com/news/
    international/21569017-artificial-cooling-makes-hot-places-bearablebut-
    worryingly-high-cost-no-sweat.

84 *at double-digit rates* https://www.washingtonpost.com/news/
    energy-environment/wp/2016/05/31/the-world-is-about-to-install-700-
    million-air-conditioners-heres-what-that-means-for-the-climate/.

84 *in the tropics* http://www.nytimes.com/2014/07/12/business/for-biggest-
    cities-of-2030-look-toward-the-tropics.html.

85 *the death rate during heatwaves* http://www.economist.com/news/
    international/21569017-artificial-cooling-makes-hot-places-bearablebut-
    worryingly-high-cost-no-sweat.

85 *reducing fights* http://journaltimes.com/news/local/violence-can-rise-with-
    the-heat-experts-say/article_d5f5f268-d911-556b-98b0-123bd9c6cc7c.
    html.

85 *maths tests* Geoffrey M. Heal and Jisung Park, 'Feeling the Heat:
    Temperature, Physiology & the Wealth of Nations', discussion paper 14-60,
    January 2014, http://live.belfercenter.org/files/dp60_heal-park.pdf

85 *24 per cent more work* http://content.time.com/time/nation/article/
    0,8599,2003081,00.html.

85 *the less productive people could be* http://www.pnas.org/content/103/10/3510.
    full.pdf.

85 *twenty-two degrees* Heal and Park, ibid.

85 *by two degrees* https://www.theguardian.com/environment/2015/oct/26/
    how-america-became-addicted-to-air-conditioning.

85 *when they leak* http://www.economist.com/news/international/21569017-
    artificial-cooling-makes-hot-places-bearablebut-worryingly-high-cost-no-
    sweat.

86 *energy consumption by 2050* https://www.theguardian.com/
    environment/2012/jul/10/climate-heat-world-air-conditioning.

15 Department Stores

87 *Selfridge's caused a sensation.* Lindy Woodhead, *Shopping, Seduction & Mr
    Selfridge* (London: Profile Books, 2007).

88 *for a few decades* Frank Trentmann, *Empire of Things* (London: Allen Lane,
    2016), p. 192.

88 *to be seen* Steven Johnson, *Wonderland* (New York: Riverhead Books, 2016); Trentmann, p. 192.

88 *the 'bargain basement'* Woodhead, ibid.

89 *'your admiration'* Harry E. Resseguie, 'Alexander Turney Stewart and the Development of the Department Store, 1823-1876', *The Business History Review*, 39(3), Autumn 1965, pp. 301–22.

90 *four thousand people an hour* Trentmann, pp. 191–7.

90 *'total shopping'* Trentmann, pp. 191–7.

90 *more time shopping than men do* The American Time Use Survey 2015, Table 1, shows that women spend an average of 53 minutes a day 'purchasing goods and services': men spend 36 minutes a day. https://www.bls.gov/news.release/pdf/atus.pdf.

90 *friendliness of sales assistants* Knowledge@Wharton '"Men Buy, Women Shop": The Sexes Have Different Priorities When Walking Down the Aisles', http://knowledge.wharton.upenn.edu/article/men-buy-women-shop-the-sexes-have-different-priorities-when-walking-down-the-aisles/.

91 *'helped emancipate women'* Woodhead, ibid.

III INVENTING NEW SYSTEMS

93 *no cause for complaint* See *Friendship Among Equals*, an official history of the ISO published in 1997, http://www.iso.org/iso/2012_friendship_among_equals.pdf.

16 The Dynamo

95 *'productivity statistics'* Robert M. Solow, 'We'd Better Watch Out', *New York Times Book Review*, 12 July 1987.

96 *'productivity paradox'* Robert Gordon, *The Rise and Fall of American Growth* (Oxford: Princeton University Press, 2016), pp. 546–7.

96 *the age of steam* The key references here are Paul David, 'The Computer and the Dynamo: An Historical Perspective', *American Economic Review*, May 1990, pp. 355–61 which popularised the parallel between computing in the late 20th century and electricity in the late 19th, and Warren Devine, 'From Shafts to Wires: Historical Perspective on Electrification', *Journal of Economic History*, 1983, pp. 347–72, which gives much more detail both about how the steam-powered and electric-powered factories worked, and on the take-up of the new technology over time.

99 *half a century old* Paul A. David and Mark Thomas, *The Economic Future in Historical Perspective* (Oxford: OUP/British Academy, 2006), pp. 134–43.

99 *change the whole system* Erik Brynjolfsson and Lorin M. Hitt, 'Beyond Computation: Information Technology, Organizational Transformation and Business Performance', *Journal of Economic Perspectives*, Fall 2000, pp. 23–48.

17 The Shipping Container

100 *around 50 per cent* World Bank: World Development Indicators 2016, http://data.worldbank.org/indicator/TG.VAL.TOTL.GD.ZS.

100 *A shipping container.* Wikipedia, *Intermodal Container*, https://en.wikipedia.org/wiki/Intermodal_container, accessed 4 July 2016.

101 *a nightmare* Maritime Cargo Transportation Conference (U.S.), *The S.S. Warrior* (Washington: National Academy of Sciences-National Research Council, 1954).

101 *might take three months* Marc Levinson, *The Box* (Oxford: Princeton University Press, 2008), Chapter 2. Also Alexander Klose, *The Container Principle* (London: MIT Press), 2015.

102 *narrow mountain roads* Levinson, pp. 129–30.

103 *to save money* Levinson, p. 38.

103 *at the same time* Levinson, p. 45.

104 *the Port Authority* Levinson, ibid. Also see G. Van Den Burg, *Containerisation: a modern transport system* (London: Hutchinson and Co., 1969).

105 *New York in the 1950s* Nuno Limao and Anthony Venables, 'Infrastructure, Geographical Disadvantage and Transport Costs', World Bank Research Paper 2257 (1999), http://siteresources.worldbank.org/EXTEXPCOMNET/Resources/2463593-1213975515123/09_Limao.pdf.

105 *$50 a tonne* So says the 'World Freight Rates Freight Calculator' at least: http://www.worldfreightrates.com/en/freight; $1500 for a container and a container can weigh more than 30 tons.

18 The Barcode

106 *recording the transaction* Margalit Fox, 'N. Joseph Woodland, Inventor of the Barcode, Dies at 91', *New York Times*, 12 December 2012, http://www.nytimes.com/2012/12/13/business/n-joseph-woodland-inventor-of-the-bar-code-dies-at-91.html?hp&_r=0.

107 *might be able to read* Charles Gerena, 'Reading Between the Lines', *Econ Focus* Q2 2014, Federal Reserve Bank of Richmond.

107 *a technological reality* Guru Madhavan, *Think Like an Engineer* (London: OneWorld, 2015).

108 *Universal Product Code, or UPC* Stephen A. Brown, *Revolution at the Checkout Counter* (Cambridge, MA: Harvard University Press, 1997).

108 *The barcode had been born.* Alistair Milne, *The Rise and Success of the Barcode: some lessons for Financial Services*, Loughborough University working paper, February 2013.

109 *a 1908 printing press* Milne, ibid.

109 *the world of the barcode* Thomas J. Holmes, 'Barcodes Lead to Frequent Deliveries and Superstores', *The RAND Journal of Economics*, 32(4), Winter 2001.

110 *closest rivals combined* National Retail Federation 2016, https://nrf.com/2016/global250-table – Wal-Mart revenues in 2014 were $486bn. Costco, Kroger, Lidl's parent group Schwarz, Tesco and Carrefour earned revenues of about $100bn each.

110 *inventory management* Emek Basker, 'The Causes and Consequences of Wal-Mart's Growth', *Journal of Economic Perspectives*, 21(3), Summer 2007.

110 *that customer is Wal-Mart* David Warsh, 'Big Box Ecology', *Economic Principals* 19 Feb 2006; Emek Basker and Van H. Pham, 'Putting a Smiley Face on the Dragon: Wal-Mart as Catalyst to U.S.-China Trade', University of Missouri-Columbia Working Paper, July 2005, http://dx.doi.org/10.2139/ssrn.765564.

110 *tattooed with a barcode* 'Barcodes', 99% Invisible Episode 108, http://99percentinvisible.org/episode/barcodes/.

19 The Cold Chain

111 *General Jorge Ubico* Dan Koeppel, *Banana: The Fate of the Fruit That Changed the World* (New York: Hudson Street Press, 2008).

112 *thirty-six years* Koeppel, ibid.

112 *banana republics* Koeppel, ibid.

112 *onward journey inland* Koeppel, ibid.

113 *beef exports could begin* Tom Jackson, *Chilled: How Refrigeration Changed the World and Might Do So Again* (London: Bloomsbury, 2015).

113 *much else besides* Jackson, ibid.

113 *before the ice melted* http://www.msthalloffame.org/frederick_mckinley_jones.htm.

114 *to solve that problem* http://www.bbc.co.uk/newsbeat/article/37306334/this-invention-by-a-british-student-could-save-millions-of-lives-across-the-world.

114 *lasts for longer still* Jackson, ibid.

114 *every week or two* Jackson, ibid.

114 *nearly nine in ten* http://www.bbc.co.uk/news/magazine-30925252.

115 *grow them in Sweden* Annika Carlson, 'Greenhouse gas emissions in the life cycle of carrots and tomatoes', IMES/EESS Report No 24, Lund University, 1997, http://ntl.bts.gov/lib/15000/15100/15145/DE97763079.pdf.

115 *raise a lamb in England* http://www.telegraph.co.uk/news/uknews/1553456/Greener-by-miles.html.

115 *millions of dollars' worth* http://www.trademap.org/Product_SelProductCountry.aspx?nvpm=1|320||||TOTAL|||2|1|1|2|1||1||.

115 *corn and cardamom* https://www.cia.gov/library/publications/the-world-factbook/geos/gt.html.

115 *enough to eat* https://www.usaid.gov/guatemala/food-assistance.

115 *110th of 138* http://www3.weforum.org/docs/GCR2016-2017/05FullReport/TheGlobalCompetitivenessReport2016-2017_FINAL.pdf.

20 Tradable Debt and the Tally Stick

117 *King Henry III* Hilary Jenkinson, 'Exchequer Tallies', *Archaeologia*, 62(2), January 1911, pp. 367–80 DOI: https://doi.org/10.1017/S0261340900008213; William N. Goetzmann and Laura Williams, 'From Tallies and Chirographs to Franklin's Printing Press at Passy', in William

N. Goetzmann and K. Geert Rouwenhorst, *The Origins of Value* (Oxford: Oxford University Press, 2005); and Felix Martin, *Money: The Unauthorised Biography* (London: Bodley Head, 2013) Chapter 1.

119 *a form of private money* David Graeber, *Debt: The first 5000 years* (London: Melville House, 2014), p. 47.

21 Billy Bookcase

121 *hates the Billy bookcase* http://www.dailymail.co.uk/news/article-2660005/ What-great-IKEA-Handyman-makes-living-building-flatpack-furniture-30-hour-dont-know-nuts-bolts.html.

121 *he'd forget it* http://www.dezeen.com/2016/03/14/ ikea-billy-bookcase-designer-gillis-lundgren-dies-aged-86/.

121 *every hundred people* http://www.adweek.com/news/advertising-branding/ billy-bookcase-stands-everything-thats-great-and-frustrating-about-ikea-173642.

121 *less than forty* http://www.bloomberg.com/news/articles/2015-10-15/ ikea-s-billy-bookcase-is-cheap-in-slovakia-while-the-u-s-price-is-surging.

122 *parts of the Billy bookcase* http://www.apartmenttherapy.com/ the-making-of-an-ikea-billy-bookcase-factory-tour-205339.

122 *ready for the trucks* http://www.nyteknik.se/automation/ bokhyllan-billy-haller-liv-i-byn-6401585.

122 *of the letter: 1952* http://www.apartmenttherapy.com/ the-making-of-an-ikea-billy-bookcase-factory-tour-205339.

122 *tens of billions* http://www.ikea.com/ms/en_JP/about_ikea/facts_and_ figures/ikea_group_stores/index.html.

122 *despite dyslexia* https://sweden.se/business/ingvar-kamprad-founder-of-ikea/.

122 *'unscrew the legs'* http://www.dezeen.com/2016/03/14/ ikea-billy-bookcase-designer-gillis-lundgren-dies-aged-86/.

122 *started to boycott him* https://sweden.se/business/ ingvar-kamprad-founder-of-ikea/.

122 *screwing on the armrests* http://www.wsj.com/articles/ ikea-cant-stop-obsessing-about-its-packaging-1434533401.

123 *knocking around* Rolf G. Larsson, 'Ikea's Almost Fabless Global Supply Chain – A Rightsourcing Strategy for Profit, Planet, and People', Chapter 3 in Yasuhiro Monden and Yoshiteru Minagawa (eds) *Lean Management of Global Supply Chain* (Singapore: World Scientific, 2015).

123 *kiln in Romania* Larsson, ibid.

123 *shelves in the store* http://highered.mheducation.com/sites/0070700893/ student_view0/ebook2/chapter1/chbody1/how_ikea_designs_its_sexy_ prices.html.

123 *production method* http://www.ikea.com/ms/en_CA/img/pdf/ Billy_Anniv_en.pdf.

123 *has only doubled* http://www.nyteknik.se/automation/ bokhyllan-billy-haller-liv-i-byn-6401585.

123 *bookcases for Ikea* ibid.

124 *by a tenth* Larsson, ibid.

124 *a flea market* https://www.theguardian.com/business/2016/mar/10/
ikea-billionaire-ingvar-kamprad-buys-his-clothes-at-second-hand-stalls.

124 *drive an old Volvo* https://sweden.se/business/ingvar-kamprad-founder-of-ikea/.

124 *something to do with it* http://www.forbes.com/sites/robertwood/2015/11/02/
how-ikea-billionaire-legally-avoided-taxes-from-1973-until-
2015/#6b2b40d91bb4.

124 *'without trying too hard'* http://www.adweek.com/news/advertising-branding/
billy-bookcase-stands-everything-thats-great-and-frustrating-about-ikea-
173642.

124 *an interesting quality: anonymity* http://news.bbc.co.uk/1/hi/8264572.stm.

124 *'to make feel high-end'* http://www.adweek.com/news/advertising-branding/
billy-bookcase-stands-everything-thats-great-and-frustrating-about-ikea-
173642.

124 *a baby-changing station* http://www.ikeahackers.net/category/billy/.

125 *'I prefer a challenge.'* http://www.dailymail.co.uk/news/article-2660005/
What-great-IKEA-Handyman-makes-living-building-flatpack-furniture-
30-hour-dont-know-nuts-bolts.html.

22 The Elevator

126 *What's going on?* This puzzle was from *Futility Closet* podcast, www.futilitycloset.
com.

126 *700,000 elevators a year* Precisely how many journeys are taken in lifts
is unclear. According to the National Elevator Industry Inc. 'Fun
Facts' sheet, the number is 18 billion a year in the US alone. Another
credible-seeming source is more bullish (Glen Pederick, 'How Vertical
Transportation is Helping Transform the City', Council on Tall
Buildings and Urban Habitat Working Paper, 2013). Pederick estimates
7 billion passenger journeys a day worldwide, although this produces
the suspiciously convenient factoid that elevators move the equivalent
of the entire world's population every day. The fact that we don't really
know just emphasises how underrated the elevator is. The statistic about
Chinese elevator installation comes from Andreas Schierenbeck, chairman
of ThyssenKrupp Elevators, quoted in *The Daily Telegraph*, 23 May 2015,
who puts the number at 700,000 a year.

126 *more than 400,000* The Skyscraper Center (Council on Tall Buildings and
Urban Habitat), http://skyscrapercenter.com/building/burj-khalifa/3 and
http://skyscrapercenter.com/building/willis-tower/169.

127 *clandestinely visit him* Eric A. Taub, 'Elevator Technology: Inspiring many
everyday leaps of faith', *New York Times*, 3 December 1998, http://www.
nytimes.com/1998/12/03/technology/elevator-technology-inspiring-many-
everyday-leaps-of-faith.html?_r=0.

127 *draft animals* http://99percentinvisible.org/episode/six-stories/.

127 *up from the mines* Ed Glaeser, *Triumph of the City* (London: Pan, 2012).

128 *the elevator brake* http://99percentinvisible.org/episode/six-stories/; Glaeser,

p. 138; Jason Goodwin, *Otis: Giving Rise to the Modern City* (Chicago: Ivan R. Dee, 2001).

129 *impossible without the elevator* David Owen, 'Green Manhattan', *The New Yorker*, 18 October 2004; Richard Florida, 'The World Is Spiky', *The Atlantic Monthly*, October 2005.

129 *safer than escalators* Nick Paumgarten, 'Up and Then Down', *The New Yorker*, 21 April 2008. Paumgarten points out that there are no truly reliable statistics on elevator accidents but they are clearly safe. About two people a month die in elevator accidents in the US, but almost invariably those people are working on elevator maintenance rather than being passengers. In any case, two people die on America's roads every half an hour – which puts the elevator fatality rate into context.

129 *the motors* Kheir Al-Kodmany, 'Tall Buildings and Elevators: A Review of Recent Technological Advances', *Buildings* 2015, 5, 1070-1104; doi:10.3390/buildings5031070.

129 *back to the building* Kheir Al-Kodmany, ibid., and Molly Miller, 'RMI Retrofits America's Favorite Skyscraper', Rocky Mountain Institute Press Release, http://www.rmi.org/RMI+Retrofits+America's+Favorite+Skyscraper.

130 *RMI* Writing in 2004, David Owen made the point that the RMI was split across two sites. More recently the RMI has opened a new headquarters.

130 *energy-saving heat exchangers* Rocky Mountain Institute Visitor's Guide: http://www.rmi.org/Content/Files/Locations_LovinsHome_Visitors_Guide_2007.pdf.

IV IDEAS ABOUT IDEAS

132 *galvanic batteries* George M. Shaw, 'Sketch of Thomas Alva Edison', *Popular Science Monthly* Vol. 13, August 1878, p. 489, https://en.wikisource.org/wiki/Popular_Science_Monthly/Volume_13/August_1878/Sketch_of_Thomas_Alva_Edison.

132 *'every six months or so'* Rutgers University Edison and Innovation Series: 'The Invention Factory', http://edison.rutgers.edu/inventionfactory.htm.

23 Cuneiform

133 *drunk himself insensible* Felix Martin, *Money: The Unauthorised Biography* (London: Bodley Head, 2013), pp. 39–42.

134 *stylised and standardised* William N. Goetzmann, *Money Changes Everything: How Finance Made Civilization Possible* (Oxford: Princeton University Press, 2016), pp. 19–30.

134 *they are the same* Jane Gleeson-White, *Double Entry: How the Merchants of Venice Created Modern Finance* (London: Allen & Unwin, 2012), pp. 11–12.

137 *compound interest* Goetzmann, ibid.

24 Public Key Cryptography

138 *to keep schtum* http://alumni.stanford.edu/get/page/magazine/article/
?article_id=74801.

139 *some predetermined number* http://www.eng.utah.edu/~nmcdonal/Tutorials/
EncryptionResearchReview.pdf.

140 *would have said it was* http://www.theatlantic.com/magazine/archive/
2002/09/a-primer-on-public-key-encryption/302574/.

141 *went unclaimed* http://www.eng.utah.edu/~nmcdonal/Tutorials/
EncryptionResearchReview.pdf.

141 *the spy chief had not* http://alumni.stanford.edu/get/page/magazine/article/
?article_id=74801.

142 *an unlikely friendship* ibid.

142 *an open book* http://www.digitaltrends.com/computing/quantum-
computing-is-a-major-threat-to-crypto-says-the-nsa/.

25 Double-Entry Bookkeeping

143 *'Draw Milan'* Robert Krulwich, 'Leonardo's To Do List', NPR, 18 November
2011, http://www.npr.org/sections/krulwich/2011/11/18/142467882/
leonardos-to-do-list.

143 *from Maestro Luca* Jane Gleeson-White, *Double Entry: How the Merchants of
Venice Created Modern Finance* (London: Allen and Unwin, 2013), p. 49.

143 *a big fan of Maestro Luca* Raffaele Pisano, 'Details on the mathematical interplay
between Leonardo da Vinci and Luca Pacioli', *BSHM Bulletin: Journal of the
British Society for the History of Mathematics* 31(2), 2016, pp. 104–111, DOI:
10.1080/17498430.2015.1091969.

144 *around 1300* Alfred W. Crosby, *The Measure of Reality: Quantification and
Western Society, 1250–1600* (Cambridge: Cambridge University Press, 1996),
Chapter 10.

144 *thousands of years* Omar Abdullah Zaid, 'Accounting Systems and Recording
Procedures in the Early Islamic State', *Accounting Historians Journal* 31(2),
December 2004, pp. 149–70, and Gleeson-White, p. 22.

144 *a purely oral tradition* Jolyon Jenkins, *A Brief History of Double Entry
Bookkeeping*, BBC Radio 4 series, March 2010, Episode 5.

144 *borrowing and lending* William N. Goetzmann, *Money Changes Everything:
How Finance Made Civilization Possible* (Woodstock: Princeton University
Press, 2016), pp. 199–201.

145 *ordered the wool* Crosby, p. 201; Crosby relies on Iris Origo, *The Merchant of
Prato* (London: Penguin, 1992).

145 *alla Veneziana* Crosby, ibid., and Origo, ibid.

145 *he wrote the book* Michael J. Fisher, 'Luca Pacioli on Business Profits', *Journal
of Business Ethics* 25, 2000, pp. 299–312.

146 *the printing industry* Gleeson-White, pp. 71–8.

146 *a Renaissance mathematician* Gleeson-White, pp. 115–120.

147 *the modern world* Anthony Hopwood, 'The archaeology of accounting

systems', *Accounting, organizations and society* 12(3), 1987, pp. 207–34; Gleeson-White, pp. 136–8; Jenkins, Episode 6.

147 *warn us of this* Gleeson-White, p. 215.

148 *in ignominy by 1850* Jenkins, Episode 7.

148 *'all such reckonings!'* Translation by Larry D Benson, https://sites.fas.harvard. edu/~chaucer/teachslf/shippar2.htm.

26 Limited Liability Companies

149 *'of modern times'* David A. Moss, *When All Else Fails: Government as the Ultimate Risk Manager* (Cambridge, MA: Harvard University Press, 2002).

149 *ancient Rome* Ulrike Malmendier, 'Law and Finance at the Origin', *Journal of Economic Literature*, 47(4), December 2009, pp. 1076–1108.

151 *'pioneers of the industrial revolution'* 'The Key to Industrial Capitalism: Limited Liability', *The Economist*, http://www.economist.com/node/347323.

151 *has its problems* Randall Morck, 'Corporations', in *The New Palgrave Dictionary of Economics*, 2nd Edition (New York: Palgrave Macmillan, 2008), Vol. 2, pp. 265–8.

151 *'watch over their own', he wrote* Adam Smith, *An Inquiry into the Nature and Causes of the Wealth of Nations*, 1776.

152 *change the law* See, for instance, Joel Bakan, *The Corporation: The Pathological Pursuit of Profit and Power* (Penguin Books Canada, 2004). An alternative view is that of the economist John Kay, who argues that Friedman's basic analysis is wrong, and there's no legal or economic reason why a corporation shouldn't pursue social goals: John Kay, 'The Role of Business in Society', https://www. johnkay.com/1998/02/03/the-role-of-business-in-society/, February 1998.

152 *American Declaration of Independence* http://www.economist.com/ node/21541753.

153 *at least five hundred* Kelly Edmiston, 'The Role of Small and Large Businesses in Economic Development', *Federal Reserve Bank of Kansas City Economic Review* Q2 2007, https://www.kansascityfed.org/PUBLICAT/ECONREV/ pdf/2q07edmi.pdf, p. 77.

153 *unfair won* http://www.pewresearch.org/fact-tank/2016/02/10/most-americans-say-u-s-economic-system-is-unfair-but-high-income-republicans-disagree/.

153 *healthy competition* http://www.economist.com/news/briefing/21695385-profits-are-too-high-america-needs-giant-dose-competition-too-much-good-thing.

27 Management Consulting

154 *unmarked piles* Slides accompanying the research paper 'Does Management Matter?', downloadable at https://people.stanford.edu/nbloom/sites/default/ files/dmm.pptx.

154 *worth their fees* Nicholas Bloom, Benn Eifert, David McKenzie, Aprajit Mahajan and John Roberts, 'Does management matter?: Evidence from India', *Quarterly Journal of Economics*, February 2013, https://people.stanford. edu/nbloom/sites/default/files/dmm.pdf.

155 *random buzzword generator* http://www.atrixnet.com/bs-generator.html.

155 *on management consultants* http://www.civilserviceworld.com/articles/news/
public-sector-spend-management-consultants-rises-second-year-row.

155 *125 billion dollars* Consultancy UK News, '10 Largest Management
Consulting Firms of the Globe', http://www.consultancy.uk/news/2149/10-
largest-management-consulting-firms-of-the-globe, 15 June 2015.

155 *over 100,000 people* Duff McDonald, 'The Making of McKinsey: A Brief
History of Management Consulting in America', *Longreads*, 23 October 2013,
https://blog.longreads.com/2013/10/23/the-making-of-mckinsey-a-brief-
history-of-management/.

156 *reviewing the past* McDonald, ibid.

156 *'you bastards!'* Hal Higdon, *The Business Healers* (New York: Random House,
1969), pp. 136–7.

157 *transformed the business world* Duff McDonald, *The Firm* (London: Simon and
Schuster, 2013).

157 *'business philosopher-kings'* Nicholas Lemann, 'The Kids in the Conference
Room', *The New Yorker*, 18 October 1999.

157 *to hire management consultants* Chris McKenna, *The World's Newest Profession*
(Cambridge University Press, 2006); see e.g. pp. 17, 21, 80.

157 *insider trading* Patricia Hurtado, 'Ex-Goldman director Rajat Gupta Back Home
After Prison Stay', *Bloomberg*, 19 January 2016, http://www.bloomberg.com/
news/articles/2016-01-19/ex-goldman-director-rajat-gupta-back-home-after-
prison-stay.

158 *Skilling went to jail* Jamie Doward, 'The Firm That Built the House of
Enron', *The Observer*, 24 March 2002, https://www.theguardian.com/
business/2002/mar/24/enron.theobserver.

158 *'land and expand'* https://www.theguardian.com/business/2016/oct/17/
management-consultants-cashing-in-austerity-public-sector-cuts.

158 *up to nine years* http://www.telegraph.co.uk/news/politics/12095961/
Whitehall-spending-on-consultants-nearly-doubles-to-1.3billion-in-three-
years...-with-47-paid-over-1000-a-day.html.

158 *Accenture's consulting fees* Slides accompanying the research paper 'Does
Management Matter?', downloadable at https://people.stanford.edu/
nbloom/sites/default/files/dmm.pptx; Bloom, Eifert, McKenzie, Mahajan
and Roberts, ibid.

28 Intellectual Property

159 *'any atrocious company?'* Letter to Henry Austin, 1 May 1942. Quoted in
'How the Dickens Controversy Changed American Publishing', Tavistock
Books blog, http://blog.tavbooks.com/?p=714.

160 *Dickens' campaign* Zorina Khan, 'Intellectual Property, History of', in
The New Palgrave Dictionary of Economics, 2nd Edition, Vol. 4 (New York:
Palgrave Macmillan, 2008).

160 *at all until 1991* Ronald V. Bettig, *Copyrighting Culture: The Political Economy
of Intellectual Property* (Oxford: Westview Press, 1996), p. 13.

161 *very modern ideas* Christopher May, 'The Venetian Moment: New Technologies, Legal Innovation and the Institutional Origins of Intellectual Property', *Prometheus*, 20(2), 2000, pp. 159-79.

161 *lobbying Parliament* Michele Boldrin and David Levine, *Against Intellectual Monopoly* (Cambridge: Cambridge University Press, 2008), http://www.dklevine.com/general/intellectual/againstfinal.htm, Chapter 1.

162 *brought into its domain* William W. Fisher III, *The Growth of Intellectual Property: A History of the Ownership of Ideas in the United States*, 1999, https://cyber.harvard.edu/people/tfisher/iphistory.pdf.

162 *Tesla would benefit from that* http://www.bloomberg.com/news/articles/2014-06-12/why-elon-musk-just-opened-teslas-patents-to-his-biggest-rivals.

163 *create new ideas* For instance, see Alex Tabarrok, 'Patent Theory vs Patent Law', *Contributions to Economic Analysis and Policy*, 1(1), 2002, https://mason.gmu.edu/~atabarro/PatentPublished.pdf.

163 *in today's terms* Dickens made £38,000. In inflation-adjusted terms that's more than £3,000,000 in today's money, and relative to the cost of labour it is nearly £25,000,000.

29 The Compiler

165 *error-prone manual labour* Kurt W. Beyer, *Grace Hopper and the Invention of the Information Age* (Cambridge, MA: MIT Press, 2009).

166 *becoming a professor instead* Lynn Gilbert and Gaylen Moore, *Particular Passions: Grace Murray Hopper*, Women of Wisdom, (New York: Lynn Gilbert Inc., 2012).

166 *a pen and paper* Beyer, ibid.

167 *'all kinds of symbols'* Gilbert and Moore, ibid.

167 *'as lazy as I was'* Gilbert and Moore, ibid.

167 *solved it in a day* Beyer, ibid.

168 *Hopper called them* Beyer, ibid.

30 The iPhone

171 *most profitable in history* 'What's the World's Most Profitable Product?', BBC World Service, 20 May 2016, http://www.bbc.co.uk/programmes/p03vqgwr.

172 *artificial intelligence agent* Mariana Mazzucato, *The Entrepreneurial State* (London: Anthem Press, 2015), p. 95 – and Chapter 5 in general.

173 *governments across Europe* Mazzucato, pp. 103–5 and 'The History of CERN', http://timeline.web.cern.ch/timelines/The-history-of-CERN?page=1.

173 *the early 1960s* Katie Hafner and Matthew Lyon, *Where Wizards Stay Up Late* (London: Simon and Schuster, 1998).

173 *in the 1980s* Greg Milner, *Pinpoint: How GPS Is Changing Technology, Culture and Our Minds* (London: W.W. Norton, 2016).

173 *testing nuclear weapons* Daniel N. Rockmore, 'The FFT – an algorithm the whole family can use', *Computing Science Engineering*, 2(1), 2000, p. 60. http://www.cs.dartmouth.edu/~rockmore/cse-fft.pdf.

173 *agency of the British government* Florence Ion, 'From touch displays to the

Surface: A brief history of touchscreen technology', *Ars Technica*, 4 April 2013, http://arstechnica.com/gadgets/2013/04/from-touch-displays-to-the-surface-a-brief-history-of-touchscreen-technology/.

174   *and the CIA* Mazzucato, pp. 100–3.

174   *an undisclosed sum* Danielle Newnham, 'The Story Behind Siri', *Medium*, 21 August 2015, https://medium.com/swlh/the-story-behind-siri-fbeb109938b0#.c3eng12zr and Mazzucato, Chapter 5.

174   *some arm of the US military* Mazzucato, Chapter 5.

174   *military procurement* William Lazonick, *Sustainable Prosperity in the New Economy?: Business Organization and High-Tech Employment in the United States* (Kalamazoo: Upjohn Press, 2009).

31 Diesel Engines

176   *an assumption* See e.g. Morton Grosser, *Diesel, the Man and the Engine*, 1978; http://www.newhistorian.com/the-mysterious-death-of-rudolf-diesel/4932/; http://www.nndb.com/people/906/000082660/.

177   *piled up in the streets* Robert J. Gordon, *The Rise and Fall of American Growth* (Oxford: Princeton University Press, 2016), p. 48.

177   *a godsend* Gordon, pp.51–2.

177   *useful work* http://auto.howstuffworks.com/diesel.htm.

177   *top 50 per cent* Vaclav Smil, 'The two prime movers of globalization: history and impact of diesel engines and gas turbines', *Journal of Global History*, 2, 2007.

178   *to cause explosions* ibid.

178   *going off accidentally* ibid.

178   *France's submarines* ibid.

178   *'to British Government'* http://www.history.com/this-day-in-history/inventor-rudolf-diesel-vanishes.

179   *the engine of global trade* http: Smil, ibid.

179   *around the world* http://www.hellenicshippingnews.com/bunker-fuels-account-for-70-of-a-vessels-voyage-operating-cost/.

179   *more slowly than it did* http://www.vaclavsmil.com/wp-content/uploads/docs/smil-article-20070000-jgh-2007.pdf.

179   *steam-powered cars* http://www.huppi.com/kangaroo/Pathdependency.htm.

180   *petroleum products* Greg Pahl, *Biodiesel: Growing a New Energy Economy* (White River Junction, VT: Chelsea Green Publishing, 2008).

180   *'Big Oil Trusts'* http://www.history.com/this-day-in-history/inventor-rudolf-diesel-vanishes.

32 Clocks

181   *faster than the original* http://www.exetermemories.co.uk/em/_churches/stjohns.php.

181   *'the railway time'* Ralph Harrington, 'Trains, technology and time-travellers: how the Victorians re-invented time', quoted in John Hassard, *The Sociology of Time* (Basingstoke: Palgrave Macmillan, 1990), p. 126.

182 *risk of collisions* Stuart Hylton, *What the Railways Did for Us* (Stroud: Amberley Publishing Limited, 2015).

182 *marks on candles* https://en.wikipedia.org/wiki/History_of_timekeeping_devices.

182 *fifteen minutes a day* http://www.historyofinformation.com/expanded.php?id=3506.

183 *a couple of seconds a day* For a discussion of innovation prizes in general, see my book *Adapt: Why Success Always Starts With Failure* (New York: Farrar Straus and Giroux/London: Little Brown, 2016); Robert Lee Hotz, 'Need a Breakthrough? Offer Prize Money', *Wall Street Journal*, 13 December 2016, http://www.wsj.com/articles/need-a-breakthrough-offer-prize-money-1481043131.

184 *three hundred million years* http://www.timeanddate.com/time/how-do-atomic-clocks-work.html and Hattie Garlick interview with Demetrios Matsakis, 'I Keep the World Running On Time', *The Financial Times*, 16 December 2016, https://www.ft.com/content/3eca8ec4-c186-11e6-9bca-2b93a6856354.

184 *professional pride* https://muse.jhu.edu/article/375792.

185 *people respond to them* http://www.theatlantic.com/business/archive/2014/04/everything-you-need-to-know-about-high-frequency-trading/360411/.

185 *communications networks* http://www.pcworld.com/article/2891892/why-computers-still-struggle-to-tell-the-time.html.

185 *the Earth's ionosphere* https://theconversation.com/sharper-gps-needs-even-more-accurate-atomic-clocks-38109.

185 *five billion years* http://www.wired.co.uk/article/most-accurate-atomic-clock-ever.

33 The Haber-Bosch Process

186 *the dawn of a new era* Daniel Charles, *Master Mind: The Rise and Fall of Fritz Haber* (New York: HarperCollins, 2005).

187 *killed herself* http://jwa.org/encyclopedia/article/immerwahr-clara

187 *would not be alive today* Vaclav Smil, *Enriching the Earth: Fritz Haber, Carl Bosch, and the Transformation of World Food Production* (Cambridge, MA: MIT Press, 2004).

187 *crop rotation* http://www.wired.com/2008/05/nitrogen-it-doe.

189 *all the world's energy* http://www.rsc.org/chemistryworld/2012/10/haber-bosch-ruthenium-catalyst-reduce-power.

189 *15 per cent* http://www.vaclavsmil.com/wp-content/uploads/docs/smil-article-worldagriculture.pdf.

189 *kill the fish below* http://www.nature.com/ngeo/journal/v1/n10/full/ngeo325.html.

189 *in the coming century* http://www.nature.com/ngeo/journal/v1/n10/full/ngeo325.html.

189 *a global experiment* Thomas Hager, *The Alchemy of Air* (New York: Broadway Books, 2009).

190 *four billion people* Hager, ibid.

190 *'I have lost.'* Charles, ibid.

34 Radar

191 *Nairobi airport* http://www.telegraph.co.uk/news/worldnews/
africaandindianocean/kenya/7612869/Iceland-volcano-As-the-dust-settles-
Kenyas-blooms-wilt.html.

191 *get on flights every day* http://www.iata.org/pressroom/pr/pages/2012-12-06-
01.aspx.

191 *five billion dollars* http://www.oxfordeconomics.com/my-oxford/projects/
129051.

193 *pursuing such an idea* Robert Buderi, *The Invention That Changed the World:
The story of radar from war to peace* (London: Little, Brown, 1997), pp. 54–6.

193 *the boat didn't* Buderi, pp. 27–33.

194 *his British colleagues* Buderi, pp. 41–6.

194 *ten Nobel laureates* Buderi, p. 48.

194 *helped to win the war* Buderi, p. 246.

194 *thirty-eight million* Buderi, p. 458.

194 *slow and patchy* Buderi, p. 459.

194 *'see and be seen'* http://www.cbsnews.com/news/1956-grand-canyon-
airplane-crash-a-game-changer/.

194 *through a cloud* http://lessonslearned.faa.gov/UAL718/CAB_accident_report.pdf.

195 *is it good enough?* https://www.washingtonpost.com/world/national-security/
faa-drone-approvals-bedeviled-by-warnings-conflict-internal-e-mails-
show/2014/12/21/69d8a07a-86c2-11e4-a702-fa31ff4ae98e_story.html.

195 *concentrated minds* http://www.cbsnews.com/news/1956-grand-canyon-
airplane-crash-a-game-changer/.

195 *in the United States* https://www.faa.gov/about/history/brief_history/.

195 *twenty times busier still* http://www.transtats.bts.gov/.

195 *nearly two a minute* http://www.iata.org/publications/Documents/iata-
safety-report-2014.pdf.

35 Batteries

197 *needed hanging again* The Newgate Calendar, http://www.exclassics.com/
newgate/ng464.htm.

198 *commercialised it* http://www.economist.com/node/10789409.

199 *thirty minutes* http://content.time.com/time/specials/packages/
article/0,28804,2023689_2023708_2023656,00.html.

199 *processing power* http://www.economist.com/news/
technology-quarterly/21651928-lithium-ion-battery-steadily-improving-
new-research-aims-turbocharge.

199 *batteries last longer* http://www.economist.com/news/
technology-quarterly/21651928-lithium-ion-battery-steadily-improving-
new-research-aims-turbocharge.

199   *all the time?* http://www.vox.com/2016/4/18/11415510/solar-power-costs-innovation.

200   *20 per cent of power* http://www.u.arizona.edu/~gowrisan/pdf_papers/renewable_intermittency.pdf.

200   *Could batteries be the solution?* http://www.bbc.co.uk/news/business-27071303.

200   *battery costs come down* http://www.rmi.org/Content/Files/RMI-TheEconomicsOfBatteryEnergyStorage-FullReport-FINAL.pdf.

200   *manufacture their 747s* http://www.fastcompany.com/3052889/elon-musk-powers-up-inside-teslas-5-billion-gigafactory.

36 Plastic

201   *north of New York City* Jeffrey L. Meikle, *American Plastic: A Cultural History* (New Brunswick: Rutgers University Press, 1995).

202   *'calls me again to rest'* Leo Baekeland, *Diary, Volume 01, 1907–1908*, Smithsonian Institute Archive Centre, https://transcription.si.edu/project/6607

202   *'It will not melt.'* Bill Laws, *Nails, Noggins and Newels* (Stroud: The History Press, 2006).

203   *196 that were* Susan Freinkel, *Plastic: A Toxic Love Story* (Boston: Houghton Mifflin Harcourt, 2011).

203   *half for energy* http://www.scientificamerican.com/article/plastic-not-so-fantastic/.

203   *'boundaries are unlimited'* Freinkel, ibid.

203   *for a bargain price?* Freinkel, ibid.

204   *phoney, superficial, ersatz* Meikle, ibid.

204   *the next twenty years* 'The New Plastics Economy: Rethinking the future of plastics', World Economic Forum, January 2016, http://www3.weforum.org/docs/WEF_The_New_Plastics_Economy.pdf.

204   *develop and reproduce* http://www.scientificamerican.com/article/plastic-not-so-fantastic/.

204   *all the fish* 'The New Plastics Economy: Rethinking the future of plastics', World Economic Forum, January 2016, http://www3.weforum.org/docs/WEF_The_New_Plastics_Economy.pdf.

204   *weigh either quantity* Leo Hornak, 'Will there be more fish or plastic in the sea by 2050?', BBC News, 15 February 2016, http://www.bbc.co.uk/news/magazine-35562253.

204   *environmental too* Richard S. Stein, 'Plastics Can Be Good for the Environment', http://www.polymerambassadors.org/Steinplasticspaper.pdf.

204   *that rate is lower still* 'The New Plastics Economy: Rethinking the future of plastics', World Economic Forum, January 2016, http://www3.weforum.org/docs/WEF_The_New_Plastics_Economy.pdf.

205   *the industry's trade association* https://en.wikipedia.org/wiki/Resin_identification_code.

205   *far from perfect* http://resource-recycling.com/node/7093.

205   *around the world* 'Environment at a Glance 2015', OECD Indicators, p. 51, http://www.keepeek.com/Digital-Asset-Management/oecd/environment/environment-at-a-glance-2015_9789264235199-en#page51.

205 *fining them if they don't* http://www.wsj.com/articles/taiwan-the-worlds-
     geniuses-of-garbage-disposal-1463519134.
205 *your 3D printer* http://www.sciencealert.com/this-new-device-recycles-
     plastic-bottles-into-3d-printing-material.
205 *with new properties* https://www.weforum.org/agenda/2015/08/turning-
     trash-into-high-end-goods/.

VI THE VISIBLE HAND

207 *'no part of his intention'* Adam Smith, *An Inquiry Into the Nature and Causes of the
     Wealth of Nations*, 1776 (pp. 455–6 of the 1976 edition, Oxford: Clarendon Press).
207 *to this day* Marc Blaug, 'Invisible Hand', in *The New Palgrave Dictionary of
     Economics*, 2nd Edition, Vol. 4 (New York: Palgrave Macmillan) 2008.

37 The Bank

209 *London's first bank* William N. Goetzmann, *Money Changes Everything: How
     Finance Made Civilization Possible* (Oxford: Princeton University Press, 2016),
     Chapter 11.
210 *operated by the government* Goetzmann, p. 180.
211 *very suspect* Fernand Braudel, *Civilization and Capitalism, 15th–18th Century:
     The structure of everyday life* (Berkeley: University of California Press, 1992),
     p. 471. The story is also told in S. Herbert Frankel's *Money: Two Philosophies*
     (Oxford: Blackwell, 1977) and by Felix Martin in *Money: The Unauthorised
     Biography* (London: Bodley Head, 2013), Chapter 6.
211 *for local currency* Martin, pp. 105–7; Marie-Thérèse Boyer-Xambeu,
     Ghislain Deleplace, Lucien Gillard and M.E. Sharpe, *Private Money & Public
     Currencies: The 16th Century Challenge* (London: Routledge, 1994).

38 Razors and Blades

215 *'endless gallery of loveliness'* King Camp Gillette, *The Human Drift* (Boston:
     New Era Publishing, 1894). Text accessed at https://archive.org/stream/
     TheHumanDrift/The_Human_Drift_djvu.txt.
215 *a year later* Randal C. Picker, 'The Razors-and-Blades Myth(s)', The Law
     School, The University of Chicago, September 2010.
216 *cheaper to produce* Picker, ibid.
216 *'quoted on this page'* Picker, ibid.
216 *got in on the act* Picker, ibid.
216 *manufacture and distribute* http://www.geek.com/games/sony-will-sell-every-
     ps4-at-a-loss-but-easily-recoup-it-in-games-ps-plus-sales-1571335/.
216 *the coffee pods* http://www.emeraldinsight.com/doi/full/
     10.1108/02756661311310431
217 *a generic cup of joe* http://www.macleans.ca/society/life/single-serve-coffee-
     wars-heat-up/.

217 *vendors offer free trials* Chris Anderson, *Free* (Random House, 2010).

217 *brand loyalty* Paul Klemperer, 'Competition when consumers have switching costs: an overview with applications to industrial organization, macroeconomics and international trade', *Review of Economic Studies*, 62, 1995.

217 *compatible blades* http://www.law.uchicago.edu/files/file/532-rcp-razors.pdf.

218 *proved difficult* For a discussion of confusion pricing, see Tim Harford, 'The Switch Doctor', *The Financial Times*, 27 April 2007, https://www.ft.com/content/921b0182-f14b-11db-838b-000b5df10621 and 'Cheap Tricks', *The Financial Times*, 16 February 2007, https://www.ft.com/content/5c15b0f4-bbf5-11db-9cbc-0000779e2340.

39 Tax Havens

219 *headquartered in Bermuda* http://www.finfacts.ie/irishfinancenews/article_1026675.shtml.

219 *completely legally* https://www.theguardian.com/business/2012/oct/21/multinational-firms-tax-ebay-ikea, http://fortune.com/2016/03/11/apple-google-taxes-eu/.

220 *they stole it from* http://www.pbs.org/wgbh/pages/frontline/shows/nazis/readings/sinister.html.

220 *lane at customs* HMRC, *Measuring Tax Gaps 2016*, https://www.gov.uk/government/uploads/system/uploads/attachment_data/file/561312/HMRC-measuring-tax-gaps-2016.pdf.

220 *Great Council of Geneva* Miroslav N. Jovanović, *The Economics of International Integration*, Second Edition (Cheltenham: Edward Elgar Publishing, 2015), p. 480.

221 *disclose financial information* Gabriel Zucman, *The Hidden Wealth of Nations: The Scourge of Tax Havens* (Chicago: University of Chicago Press, 2015).

221 *leaked away to these islands* Daniel Davies, 'Gaps and Holes: How the Swiss Cheese Was Made', *Crooked Timber Blog*, 8 April 2016, http://crookedtimber.org/2016/04/08/gaps-and-holes-how-the-swiss-cheese-was-made/.

222 *handling such questions* Gabriel Zucman, 'Taxing across Borders: Tracking Personal Wealth and Corporate Profits', *Journal of Economic Perspectives*, 28(4), Fall 2014, pp. 121–48.

222 *8500 dollars apiece* Nicholas Shaxson, *Treasure Islands: Tax Havens and the Men who Stole the World* (London: Vintage Books, 2011).

222 *foreign aid* Global Financial Integrity (GFI) programme at the Center for International Policy in Washington estimates, quoted in Shaxson, ibid.

222 *individual states* Zucman, ibid.

223 *lacked teeth* Zucman, ibid.

40 Leaded Petrol

224 *forty-nine people worked there* Gerald Markowitz and David Rosner, *Deceit and Denial: The Deadly Politics of Industrial Pollution* (Berkeley: University of California Press, 2013).

225 *'the loony gas building'* http://www.wired.com/2013/01/looney-gas-and-lead-poisoning-a-short-sad-history/.

225 *after two deaths* William J. (Bill) Kovarik, 'The Ethyl Controversy: How the news media set the agenda for a public health controversy over leaded gasoline, 1924-1926', Ph.D. Dissertation, University of Maryland DAI 1994 55(4): 781-782-A. DA9425070.

225 *'the house of butterflies'* http://pittmed.health.pitt.edu/jan_2001/butterflies.pdf.

225 *approve the findings* Markowitz and Rosner, ibid.

226 *'essential in our civilization'* Markowitz and Rosner, ibid.

226 *'Blindness, Stupidity'* Kassia St Clair, *The Secret Lives of Colour* (London: John Murray, 2016).

226 *'of a pallid colour'* http://penelope.uchicago.edu/~grout/encyclopaedia_romana/wine/leadpoisoning.html.

227 *unleaded petrol* Jessica Wolpaw Reyes, 'Environmental policy as social policy? The impact of childhood lead exposure on crime', NBERWorking Paper 13097, 2007, http://www.nber.org/papers/w13097.

227 *'environmental Kuznets curve'* I wrote about the environmental Kuznets curve in China: Tim Harford, 'Hidden Truths Behind China's Smokescreen', *Financial Times*, 29 January 2016, https://www.ft.com/content/4814ae2c-c481-11e5-b3b1-7b2481276e45.

227 *unless you drank it* http://www.thenation.com/article/secret-history-lead/.

227 *Tetraethyl lead could,* ibid.

227 *the cost of all the crime* Wolpaw Reyes, ibid.

227 *learning less in school* http://www.cdc.gov/nceh/lead/publications/books/plpyc/chapter2.htm.

227 *you could tell about asbestos* http://www.nature.com/nature/journal/v468/n7326/full/468868a.html.

227 *or tobacco* http://www.ucdmc.ucdavis.edu/welcome/features/20071114_cardio-tobacco/.

227 *General Motors* Markowitz and Rosner, ibid.

227 *conflicts of interest in research* https://www.ncbi.nlm.nih.gov/books/NBK22932/#_a2001902bddd00028._

41 Antibiotics in Farming

230 *stop them getting sick* Philip Lymbery and Isabel Oakeshott, *Farmageddon: The true cost of cheap meat* (London: Bloomsbury, 2014), pp. 306–7.

230 *keep disease in check* http://www.bbc.co.uk/news/health-35030262.

230 *fatter animals* http://www.scientificamerican.com/article/antibiotics-linked-weight-gain-mice/.

230 *sick humans* 'Antimicrobials in agriculture and the environment: Reducing unnecessary use and waste', The Review on Antimicrobial Resistance Chaired by Jim O'Neill, December 2015.

230 *to double in twenty years,* ibid.

230 *they should know better* http://ideas.time.com/2012/04/16/why-doctors-uselessly-prescribe-antibiotics-for-a-common-cold/.

230 *over the counter* http://cid.oxfordjournals.org/content/48/10/1345.full.

230 *a hundred trillion dollars* 'Antimicrobial Resistance: Tackling a crisis for the

health and wealth of nations', The Review on Antimicrobial Resistance Chaired by Jim O'Neill, December 2014.

231 *became a bacteriologist* http://www.pbs.org/wgbh/aso/databank/entries/bmflem.html.

231 *used the dish to cultivate* http://time.com/4049403/alexander-fleming-history/.

231 *nobody paid attention* http://www.nobelprize.org/nobel_prizes/medicine/laureates/1945/fleming-lecture.pdf.

231 *Fleming's old article* http://www.abc.net.au/science/slab/florey/story.htm.

231 *'penicillin girls'* http://news.bbc.co.uk/local/oxford/hi/people_and_places/history/newsid_8828000/8828836.stm; https://www.biochemistry.org/Portals/0/Education/Docs/Paul%20brack.pdf; http://www.ox.ac.uk/news/science-blog/penicillin-oxford-story.

232 *'not sufficient to kill them'* http://www.nobelprize.org/nobel_prizes/medicine/laureates/1945/fleming-lecture.pdf.

232 *pipeline dried up* http://www3.weforum.org/docs/WEF_GlobalRisks_Report_2013.pdf.

232 *compounds in soil* http://phenomena.nationalgeographic.com/2015/01/07/antibiotic-resistance-teixobactin/.

233 *antibiotic use in pigs* 'Antimicrobials in agriculture and the environment: Reducing unnecessary use and waste', The Review on Antimicrobial Resistance Chaired by Jim O'Neill, December 2015.

42 M-Pesa

235 *collect your salaries myself* http://www.technologyreview.es/printer_friendly_article.aspx?id=39828.

236 *back home in the village* Nick Hughes and Susie Lonie, 'M-Pesa: Mobile Money for the "Unbanked" Turning Cellphones into 24-Hour Tellers in Kenya', *innovations*, Winter & Spring 2007, http://www.gsma.com/mobilefordevelopment/wp-content/uploads/2012/06/innovationsarticleonmpesa_0_d_14.pdf.

237 *kiosks in Kenya as ATMs* Isaac Mbiti and David N. Weil, 'Mobile Banking: The Impact of M-Pesa in Kenya', NBER, Working Paper 17129, Cambridge MA, June 2011 http://www.nber.org/papers/w17129.

237 *mobile money* http://www.worldbank.org/en/programs/globalfindex/overview.

237 *within a few years* http://www.slate.com/blogs/future_tense/2012/02/27/m_pesa_ict4d_and_mobile_banking_for_the_poor_.html.

237 *developing country markets* http://www.forbes.com/sites/danielrunde/2015/08/12/m-pesa-and-the-rise-of-the-global-mobile-money-market/#193f89d23f5d.

237 *banking and telecoms regulators* http://www.economist.com/blogs/economist-explains/2013/05/economist-explains-18.

237 *as forthcoming* http://www.forbes.com/sites/danielrunde/2015/08/12/m-pesa-and-the-rise-of-the-global-mobile-money-market/#193f89d23f5d.

237 *sending money home* http://www.cgap.org/sites/default/files/CGAP-Brief-Poor-People-Using-Mobile-Financial-Services-Observations-on-Customer-Usage-and-Impact-from-M-PESA-Aug-2009.pdf.

238 *used as evidence* http://www.bloomberg.com/news/articles/2014-06-05/
safaricoms-m-pesa-turns-kenya-into-a-mobile-payment-paradise.

238 *a quarter of GDP* http://www.spiegel.de/international/world/corruption-in-
afghanistan-un-report-claims-bribes-equal-to-quarter-of-gdp-a-672828.html.

238 *theft and extortion* http://www.coastweek.com/3745-Transport-reolution-
Kenya-minibus-operators-launch-cashless-fares.htm.

238 *isn't hard to work out* http://www.iafrikan.com/2016/09/21/kenyas-cashless-
payment-system-was-doomed-by-a-series-of-experience-design-failures/.

43 Property Registers

239 *rice fields of Bali, Indonesia* Hernando de Soto, *The Mystery of Capital* (New
York: Basic Books, 2000), p. 163.

240 *three attempts to kill him* 'The Economist versus The Terrorist, *The Economist*,
30 January 2003, http://www.economist.com/node/1559905.

240 *owned by the collective* David Kestenbaum and Jacob Goldstein, 'The Secret
Document That Transformed China', NPR *Planet Money*, 20 January
2012, http://www.npr.org/sections/money/2012/01/20/145360447/
the-secret-document-that-transformed-china.

242 *clearly a huge amount* Christopher Woodruff, 'Review of de Soto's *The Mystery of
Capital*', *Journal of Economic Literature*, 39, December 2001, pp. 1215–23.

242 *cadastral maps there, as well* World Bank, *Doing Business in 2005* (Washington
DC: The World Bank Group, 2004), p. 33.

242 *claims on the land* Robert Home and Hilary Lim, *Demystifying the Mystery of
Capital: land tenure and poverty in Africa and the Caribbean* (London: Glasshouse
Press, 2004), p. 17.

242 *of much significance* Home and Lim, pp. 12–13; Soto, pp. 105–52.

243 *too time consuming* de Soto, pp. 20–1; World Bank, ibid.

243 *more in their land* Tim Besley, 'Property Rights and Investment Incentives:
Theory and Evidence From Ghana', *Journal of Political Economy*, 103(5),
October 1995, pp. 903–37.

243 *more private investment* World Bank, ibid.

44 Paper

247 *just twenty-eight* Mark Kurlansky, *Paper: Paging Through History* (New York:
W.W. Norton, 2016), pp. 104–5.

248 *and North Africa* Jonathan Bloom, *Paper Before Print* (New Haven: Yale
University Press, 2001).

248 *and much else* James Moseley, 'The Technologies of Print', in M.F. Suarez,
S.J. and H.R. Woudhuysen, *The Book: A Global History* (Oxford: Oxford
University Press, 2013).

248 *the skins of 250 sheep* Kurlansky, p. 82.

249 *the start of paper's uses* Mark Miodownik, *Stuff Matters* (London: Penguin,
2014), Chapter 2.

249 *a humanitarian emergency* Kurlansky, p. 46.

249 *massive drop-hammers* Kurlansky, pp. 78–82.

249 *a strong, flexible mat* Miodownik, ibid.

250 *world's first daily newspaper* Kurlansky, p. 204.

250 *to sell to paper mills* Kurlansky, p. 244.

250 *paper production in the West* Bloom, Chapter 1.

250 *turned into another box* Kurlansky, p. 295.

251 *weak and unusable* 'Cardboard', *Surprisingly Awesome 19*, Gimlet Media, August 2016, https://gimletmedia.com/episode/19-cardboard/.

251 *quarter of a century* Abigail Sellen and Richard Harper, *The Myth of the Paperless Office* (Cambridge, MA: MIT, 2001).

251 *every five years* The estimate, from Hewlett Packard in 1996, was that enough office paper emerged from printers and copiers that year to cover 18 per cent of the surface area of the US (Bloom, Ch. 1). Office paper consumption continued to rise over the subsequent few years.

251 *starting to decline* 'World wood production up for fourth year; paper stagnant as electronic publishing grows', UN Press Release 18 December 2014, http://www.un.org/apps/news/story.asp?NewsID=49643#.V-T2S_ArKUn.

252 *more bicycles than cars* David Edgerton, *Shock Of The Old: Technology and Global History since 1900* (London: Profile, 2008).

45 Index Funds

253 *'S&P 500 index fund'* NPR *Planet Money*, 'Brilliant vs Boring', 4 May 2016, http://www.npr.org/sections/money/2016/03/04/469247400/episode-688-brilliant-vs-boring

254 *the Wall Street Journal* Pierre-Cyrille Hautcoeur, 'The Early History of Stock Market Indices, with Special Reference to the French Case', Paris School of Economics working paper, http://www.parisschoolofeconomics.com/hautcoeur-pierre-cyrille/Indices_anciens.pdf.

254 *prizes in economics* Michael Weinstein, 'Paul Samuelson, Economist, Dies at 94', *New York Times*, 13 December 2009, http://www.nytimes.com/2009/12/14/business/economy/14samuelson.html?pagewanted=all&_r=0.

255 *investors to rush in* John C. Bogle, 'How the Index Fund Was Born', *The Wall Street Journal*, 3 September 2011, http://www.wsj.com/articles/SB10001424053111904583204576544681577401622.

256 *active stock-pickers* Robin Wigglesworth and Stephen Foley, 'Active asset managers knocked by shift to passive strategies', *Financial Times*, 11 April 2016, https://www.ft.com/content/2e975946-fdbf-11e5-b5f5-070dca6d0a0d.

256 *how to value them* Donald MacKenzie, 'Is Economics Performative? Option Theory and the Construction of Derivatives Markets', http://www.lse.ac.uk/accounting/CARR/pdf/MacKenzie.pdf.

257 *financial markets have moved* Brian Wesbury and Robert Stein, 'Why mark-to-market accounting rules must die', *Forbes*, 23 February 2009, http://www.forbes.com/2009/02/23/mark-to-market-opinions-columnists_recovery_stimulus.html.

257 *hundreds of billions of dollars* Eric Balchunas, 'How the Vanguard Effect Adds Up to $1 Trillion', *Bloomberg* 30 August 2016, https://www.bloomberg.com/view/articles/2016-08-30/how-much-has-vanguard-saved-investors-try-1-trillion.

257 *Something new under the sun.* Paul Samuelson speech to Boston Security
Analysts Society on 15 November 2005, cited in Bogle, ibid.

46 The S-Bend

258 *'It stinks.'* http://www.thetimes.co.uk/tto/law/columnists/article2047259.ece.
258 *encouraging sewage into gullies* G.C. Cook, 'Construction of London's Victorian
sewers: the vital role of Joseph Bazalgette', *Postgraduate Medical Journal*, 2001.
259 *'the folly of our carelessness'* Stephen Halliday, *The Great Stink of London: Sir
Joseph Bazalgette and the Cleansing of the Victorian Metropolis* (Stroud: The
History Press, 2013).
260 *Crystal Palace* Laura Perdew, *How the Toilet Changed History* (Minneapolis:
Abdo Publishing, 2015).
261 *a big step forward* Johan Norberg, *Progress: Ten Reasons to Look Forward to the
Future* (London: OneWorld, 2016) p. 33.
261 *sanitation systems that do that* http://pubs.acs.org/doi/abs/10.1021/es304284f.
261 *in Cambodia, 7 per cent* http://www.wsp.org/sites/wsp.org/files/publications/WSP-
ESI-Flier.pdf; http://www.wsp.org/content/africa-economic-impacts-sanitation;
http://www.wsp.org/content/south-asia-economic-impacts-sanitation.
262 *would appreciate it, too* 'Tackling the Flying Toilets of Kibera', Al Jazeera, http://
www.aljazeera.com/indepth/features/2013/01/201311810421796400.html 22
January 2013; Cyrus Kinyungu 'Kibera's Flying Toilets Flushed Out by PeePoo
Bags' http://bhekisisa.org/article/2016-05-03-kiberas-flying-toilets-flushed-
out-by-peepoo-bags.
262 *a flushing toilet* http://www.un.org/apps/news/story.asp?NewsID=44452#.
VzCnKPmDFBc.
262 *the constraints of Kibera* 'World Toilet Day: Kibera Slum Hopes
to Ground "Flying Toilets"', http://www.dw.com/en/
world-toilet-day-kibera-slum-seeks-to-ground-flying-toilets/a-18072068.
262 *cubic metres of earth* http://www.bbc.co.uk/england/sevenwonders/london/
sewers_mm/index.shtml.
262 *Less than 6 per cent.* G.R.K. Reddy, *Smart and Human: Building Cities of
Wisdom* (HarperCollins Publishers India, 2015).
263 *'a handkerchief to his nose'* Halliday, ibid.

47 Paper Money

265 *coins made of iron* William N. Goetzmann, *Money Changes Everything: How
Finance Made Civilization Possible* (Woodstock: Princeton University Press,
2016), Chapter 9.
265 *body weight in iron coins* William N. Goetzmann and K. Geert Rouwenhorst,
The Origins of Value (Oxford: Oxford University Press, 2005), p. 67; also
Glyn Davies, *History of Money: From Ancient Times to the Present Day* (Cardiff:
University of Wales Press, 2010), pp. 180–3.
267 *not when you left* There's a more detailed discussion of hyperinflation in my
book *The Undercover Economist Strikes Bank* (New York: Riverhead/London:
Little Brown, 2013).

48 Concrete

269 *right into the living room* M.D. Cattaneo, S. Galiani, P.J. Gertler, S. Martinez and R. Titiunik, 'Housing, health, and happiness', *American Economic Journal: Economic Policy*, 2009, pp. 75–105; Charles Kenny, 'Paving Paradise' *Foreign Policy* 3 January 2012, http://foreignpolicy.com/2012/01/03/paving-paradise/.

270 *as much greenhouse gas as aviation* Vaclav Smil, *Making the Modern World: Materials and Dematerialization* (Chichester: Wiley, 2013), pp. 54–7.

270 *'Yes and No.'* Adrian Forty, *Concrete and Culture* (London: Reaktion Books, 2012), p. 10.

270 *and therefore concrete* Nick Gromicko and Kenton Shepard, 'The History of Concrete', https://www.nachi.org/history-of-concrete.htm#ixzz31V47Zuuj; Adam Davidson and Adam McKay, 'Surprisingly Awesome: Concrete', 17 November 2015, https://gimletmedia.com/episode/3-concrete/; Amelia Sparavigna, 'Ancient Concrete Works', Working Paper, Department of Physics, Turin Polytechnic https://arxiv.org/pdf/1110.5230.

271 *arguably until 1881* https://en.wikipedia.org/wiki/List_of_largest_domes – Brunelleschi's great dome in Florence is octagonal so the distance across varies.

271 *concrete can't* Stewart Brand, *How Buildings Learn: What Happens After They're Built* (London: Weidenfeld & Nicolson, 1997).

272 *'for the whole country'* Forty, ibid., pp. 150–5.

272 *It worked brilliantly. Inventors and Inventions* (New York: Marshall Cavendish, 2008).

272 *a perfect pairing* Mark Miodownik, *Stuff Matters* (Penguin, 2014), Chapter 3.

273 *a hundred and thirty years later* Miodownik, Chapter 3; Smil, pp. 54–7.

274 *coming out of car exhausts* 'Concrete Possibilities', *The Economist*, 21 September 2006, http://www.economist.com/node/7904224; Miodownik, p. 83; Jon Cartwright, 'The Concrete Answer to Pollution', *Horizon Magazine*, 18 December 2014, http://horizon-magazine.eu/article/concrete-answer-pollution_en.html.

274 *rewards will be high* James Mitchell Crow, 'The Concrete Conundrum', *Chemistry World*, March 2008, p. 62, http://www.rsc.org/images/Construction_tcm18-114530.pdf; Vaclav Smil (ibid., p. 98) reports that the energy used to make a tonne of steel is typically around four times the energy used to make a tonne of cement; a tonne of cement itself can be used to make several tonnes of concrete. Cement production also emits carbon dioxide independently of the energy input.

274 *the wages of farm workers* Shahidur Khandker, Zaid Bakht and Gayatri Koolwal, 'The Poverty Impact of Rural Roads: Evidence from Bangladesh', World Bank Policy Research Working Paper 3875, April 2006, http://www-wds.worldbank.org/external/default/WDSContentServer/IW3P/IB/2006/03/29/000012009_20060329093100/Rendered/PDF/wps38750rev0pdf.pdf.

49 Insurance

275 *Buddha refused to play* 'Brahmajāla Sutta: The All-embracing Net of Views',

translated by Bhikkhu Bodhi, Section 2-14, http://www.accesstoinsight. org/tipitaka/dn/dn.01.0.bodh.html.

276 *did not have to be repaid* Peter Bernstein, *Against the Gods: The Remarkable Story of Risk* (Chichester: John Wiley & Sons, 1998), p. 92.

276 *many different merchants* Swiss Re, *A History of Insurance in China*, http://media. 150.swissre.com/documents/150Y_Markt_Broschuere_China_Inhalt.pdf.

277 *famous names in insurance* Raymond Flower and Michael Wynn Jones, *Lloyd's of London* (London: David & Charles, 1974); also see Bernstein, pp. 88–91.

277 *the world's great insurance companies* Michel Albert, *Capitalism against Capitalism*, translated by Paul Haviland (London: Whurr, 1993), Chapter 5; also see John Kay, *Other People's Money* (London: Profile, 2015), pp. 61–3.

278 *much in demand* James Poterba, 'Annuities in Early Modern Europe', in William N. Goetzmann and K. Geert Rouwenhorst, *The Origins of Value* (Oxford: Oxford University Press, 2005).

278 *expanded their businesses* The study is nicely summarised by Robert Smith of NPR here: http://www.npr.org/2016/09/09/493228710/what-keeps-poor-farmers-poor, and the original paper is Dean Karlan, Robert Osei, Isaac Osei-Akoto and Christopher Udry, 'Agricultural Decisions After Relaxing Credit and Risk Constraints', *Quarterly Journal of Economics*, 2014, pp. 597–652. doi:10.1093/qje/qju002.

279 *That story did not end well.* Kay, p. 120.

CONCLUSION: LOOKING FORWARD

280 *it rose above seventy* https://ourworldindata.org/grapher/life-expectancy-globally-since-1770.

280 *any baby born in 1900* http://charleskenny.blogs.com/weblog/2009/06/the-success-of-development.html.

280 *10 per cent today* https://ourworldindata.org/grapher/world-population-in-extreme-poverty-absolute?%2Flatest=undefined&stackMode=relative.

281 *lit by artificial moons* Herman Kahn and Anthony J. Wiener, *The Year 2000. A Framework for Speculation on the Next Thirty-Three Years* (New York: Macmillan, 1967); Douglas Martin, 'Anthony J. Wiener, Forecaster of the Future, Is Dead at 81', *New York Times*, 26 June 2012, http://www.nytimes. com/2012/06/27/us/anthony-j-wiener-forecaster-of-the-future-is-dead-at-81. html.

282 *times of austerity* See e.g. http://news.mit.edu/2015/mit-report-benefits-investment-basic-research-0427; https://www.chemistryworld.com/news/randd-share-for-basic-research-in-china-dwindles/7726.article; http://www.sciencemag.org/news/2014/10/european-scientists-ask-governments-boost-basic-research.

284 *innovative new medicines* https://www.weforum.org/agenda/2016/12/how-do-we-stop-tech-being-turned-into-weapons.

285 *American truck drivers* Olivia Solon, 'Self-driving trucks: what's the future for

America's 3.5 million truckers?', *The Guardian*, 17 June 2016, https://www.theguardian.com/technology/2016/jun/17/self-driving-trucks-impact-on-drivers-jobs-us.

285 *in the United States* http://mashable.com/2006/07/11/myspace-americas-number-one/#nseApOVC85q9.

285 *the top thousand* http://www.alexa.com/siteinfo/myspace.com, accessed 20 January 2017.

286 *resistant to change* For an argument that QWERTY is a good keyboard design, and not an example of technological lock-in, see Stan Liebowitz and Stephen Margolis, 'The Fable of the Keys', *The Journal of Law and Economics*, April 1990.

Epilogue: 50 The Light Bulb

287 *brightly and controllably* William D. Nordhaus, 'Do real-output and real-wage measures capture reality? The history of lighting suggests not' in Timothy F. Bresnahan and Robert J. Gordon (eds), *The Economics of New Goods* (Chicago: University of Chicago Press, 1996), pp. 27–70; for other accounts of Nordhaus's calculations see Tim Harford, *The Logic of Life* (London: Little Brown, 2008), Steven Johnson, *How We Got To Now* (London: Particular Books, 2014) and David Kestenbaum, 'The History of Light, in 6 Minutes and 47 Seconds', NPR: *All Things Considered*, 2 May 2014, http://www.npr.org/2014/05/02/309040279/in-4-000-years-one-thing-hasnt-changed-it-takes-time-to-buy-light.

289 *richest men in the world* An excellent starting point for such calculations is the 'Measuring Worth' website, www.measuringworth.com. For Timothy Taylor's perspective see NPR's Planet Money, 12 October 2010, http://www.npr.org/sections/money/2010/10/12/130512149/the-tuesday-podcast-would-you-rather-be-middle-class-now-or-rich-in-1900.

290 *78 pounds of tallow candles* Marshall B. Davidson, 'Early American Lighting', *The Metropolitan Museum of Art Bulletin*, New Series, 3(1), Summer 1944, pp. 30–40.

291 *in today's money* Steven Johnson, *How We Got To Now* (London: Particular Books, 2014), p. 165 and Davidson, ibid.

291 *from extinction* Jane Brox, *Brilliant: The Evolution of Artificial Light* (London: Souvenir Press, 2011).

291 *cheaper and cheaper* Haitz's law: https://en.wikipedia.org/wiki/Haitz%27s_law.

291 *leave a child alone with one* Brox, p. 117; also Robert J. Gordon, *The Rise and Fall of American Growth* (Oxford: Princeton University Press, 2016), Chapter 4.

Acknowledgements

What are the fifty most interesting, engaging or counterintuitive inventions that you can think of? There was a time, a year or so ago, when I was asking this question of everyone I met – so I confess a certain paranoia that someone suggested something splendid, and I have forgotten to give them credit. I beg forgiveness in anticipation.

No danger of forgetting the contributions of Philip Ball, David Bodanis, Dominic Camus, Patricia Fara, Claudia Goldin, Charles Kenny, Armand Leroi, Mark Lynas, Arthur I. Miller, Katharina Rietzler, Martin Sandbu and Simon Singh – thank you for your wisdom and generosity.

I also owe a great debt to the economic historians, technology mavens and brilliant writers who have in various ways inspired or informed the research in this book. The references to the book tell the full story but a few names spring to mind: William N. Goetzmann, Robert Gordon, Steven Johnson, Marc Levinson, Felix Martin, Mariana Mazzucato, William Nordhaus, and the crew at some of my favourite podcasts: 99% Invisible, Planet Money, Radiolab and Surprisingly Awesome.

At Little, Brown, Tim Whiting and particularly Nithya Rae coped superbly with tight deadlines and late copy – as did the indefatigable Jake Morrissey at Riverhead. I know that many other people at Little, Brown, Riverhead and my publishers around the world are involved in getting this book into your hands – but I want to particularly thank Katie Freeman for being such a glorious nerd about concrete. My agents Sue Ayton, Helen Purvis, Zoe Pagnamenta and especially Sally Holloway juggled the complexities of the project with their usual tact and skill.

At the BBC, Rich Knight had faith in this idea from the start and Mary Hockaday was quick to commission. Ben Crighton has been an inspiring and subtle producer, ably supported by studio genius James Beard, wordsmith Jennifer Clarke, production coordinator Janet Staples, editor Richard Vadon, and many others.

As always I'm grateful to my editors at the *Financial Times* for their support and indulgence – Esther Bintliff, Caroline Daniel, Alice Fishburn, Alec Russell, Fred Studemann and others. What wonderful colleagues you are.

But the most important collaborator on this project has been Andrew Wright. Andrew researched vast swathes of this book; he also produced wise and witty first drafts of many of the chapters, as well as improving the others with his customarily incisive edits. I am grateful for his speed, skill, and self-effacing insistence that writing half a book is no big deal. I'm even more grateful for his friendship for the last quarter of a century.

Finally – thank you to my family, Fran, Stella, Africa and Herbie. You guys are awesome.

Index

To buy any of our books and to find out
more about Abacus and Little, Brown, our authors
and titles, as well as events and book clubs,
visit our website

www.littlebrown.co.uk

and follow us on Twitter

@AbacusBooks
@LittleBrownUK